SOCIETY AND SCIENCE

Decision-Making Episodes for Exploring Society, Science, and Technology

Nancy N. Stahl
Gifted Education Specialist
Arizona Department of Education
Phoenix, Arizona

Robert J. Stahl
Division of Curriculum and Instruction
Arizona State University
Tempe, Arizona

With a Foreword by Robert E. Yager

Dale Seymour Publications
Parsippany, New Jersey

Project Editor: Mali Apple
Production/Mfg. Coordinator: Leanne Collins
Design Manager: Jeff Kelly
Text and Cover Design: Lisa Raine
Text Illustrations: Rachel Gage
Cover Photo: ©1994 The Image Bank/Joseph Drivas

Dale Seymour Publications®
An imprint of Pearson Learning
299 Jefferson Road, P.O. Box 480
Parsippany, New Jersey 07054-0480
www.pearsonlearning.com
1-800-321-3106

ISBN 0-201-49097-8

3 4 5 6 7 8 9 10-ML-04 03 02 01 00

CONTENTS

The Decisions Episodes

Episodes marked with an asterisk are accompanied by a sample lesson plan in the Appendix.

Forced-Choice Decision Episodes

Rank-Order Decision Episodes

Negotiation Decision Episodes

Dedication

Individually and collectively, we dedicate this book to J. Doyle Casteel, friend, scholar, mentor, colleague. Our lives and countless others will always be so much better because of the way you touch minds, hearts, and lives.

Acknowledgments

Very special people helped us at critical periods of our exploration of society-science-technology issues and in the development of one or more of these student episodes. Marietta Franklin, Sally Collins, Patsy Goodwin, and Betsy Beals were very helpful to Nancy during her teaching days at the Ravenscroft School in Raleigh, North Carolina. For Bob, Paul Becht, Science Department, P. K. Yonge Laboratory School, University of Florida, Gainesville; Tom and Sandy Gadsden, Richardson, Texas, former colleagues at the Laboratory School, with Tom now head of the Office of Education, Superconducting Supercollider Laboratory in Dallas; John Koran, Science Education, University of Florida; and Ned Bingham, Professor Emeritus, Science Education, University of Florida, were instrumental in their support and in sharing their science and instructional expertise.

Barry Fraser, David Tresgust, Ken Tobin, and John Malone, Science and Mathematics Education Center, Curtin University of Technology, Perth, Western Australia, provided Bob opportunities to share this approach and some of these materials with Australian teachers and students.

Doyle Casteel, University of Florida, and Bob Yager, University of Iowa, who were briefly colleagues at Iowa, also shaped our thinking about the value of an integrative study of society, science, and technology issues. We want to take this opportunity to publicly thank these individuals for their positive contributions to our professional and personal lives. Finally, we appreciate the assistance of Mali Apple, our editor in this project. Her suggestions and expertise have made the enclosed highly inviting and usable for teachers and students alike.

FOREWORD

Why *Society and Science* Is Needed in YOUR Classroom

A basic problem in education is the erroneous belief that learning can occur as a result of direct teaching. We have often sought reform by structuring new curriculum frameworks, new courses of study, new textbooks. We have assumed that learning can easily be accomplished by a simple act of transmitting what teachers know or what printed materials (e.g., textbooks) report. We have acted as though students who learn the most and best do so by merely listening and reading carefully and then committing the language used and the skills stressed to memory. Most teachers and textbook tests require only recall. Many standardized achievement tests require similar tasks.

As we speed toward a new century, the results of our past failures in education have intensified and demands for changes have emerged as a top national priority (as evidenced by the President's, governors', and Congressionally endorsed education goals). The most important development may be the widespread acceptance of the necessity of reform. Everyone is united in advocating reform, restructuring, and improving schooling and formal education. However, there are widely disparate views as to what form these efforts need to take.

One of the most exciting developments of the last decade was the decision to provide funds for basic research about human learning. National Science Foundation funds, newly available in 1983, were to provide invaluable information that could be used to promote science education reform. Cognitive scientists were chosen to conduct experiments and observations designed to shed more light on how humans think, process information, and learn. After concentrating on how the most successful minds functioned (with the naive notion that this would help educators work with the less successful—thereby allowing teachers to substantially improve achievement for all students), we were astounded to find that there were few really "successful" students and adults in terms of *real learning*. Real learning was defined as the use of targeted concepts and skills applied in new situations and to new problems. For example, up to 90 percent of the undergraduate physics and engineering students at prestigious colleges and universities were unable to solve real-world problems that required use of the concepts and skills taught—and presumably learned.

These astounding results led researchers to explore *constructivism* as a viable perspective on human cognition and learning. This research has been traced back over three hundred years to Vico, an Italian philosopher. Basically, the constructivists observed individuals and conducted research that indicated that all human learning comes from engagement of the mind in ways that result in meaning being invented, developed, assigned, modified, and organized internally by each individual. Among the findings of the researchers is that many of the most able science students could not solve problems because of existing self-constructed conceptual frameworks, often consisting of explanations formulated as students' *made sense* of their experiences outside of schools. These self-generated conceptions and explanations tended to be used in real-world problem-solving situations instead of the concepts and skills students were taught in classrooms and laboratories and that their teachers assumed would be applied.

All of this new research and focus for current reform efforts provides a basis for Nancy and Bob Stahl's new book. *Society and Science* provides a perspective for the international STS movement that embraces the Constructivist Learning Theory. In their opening chapter, the Stahls have chosen to order the words from the traditional Science-Technology-Society (STS) to Society, Science, and Technology (SST) to emphasize the centrality of humans as science and technology are considered. In addition, this order reminds us that humans, individually and collectively, invent and decide on the uses of science and technology, and are affected by these decisions and uses. Both science and technology are creations of the human mind. As with STS, the question might be raised concerning the appropriateness of listing "science" before "technology." In both instances, placing science before technology does not mean that technology is always an extension of the human mind to apply science to better one's life or to fulfill personal needs or desires. Today as always technology often precedes science and is often invented or used in the absence of or in spite of science.

Society and Science looks broadly at teaching and learning and includes a rationale and examples of how the two must be interrelated in classrooms. Learning surely should be a result, not just an expectation, of teaching. However, in practice, typical teaching often results in little or no learning, as evidenced by what elementary through twelfth grade level science students and undergraduate physics and engineering majors can actually *do* when faced with a relevant problem to solve outside the formal laboratory. Too few see the real laboratory (a place to investigate) as the whole universe. Instead, a laboratory is perceived as a place students go to work on experiments conducted following clear directions and with preferred results known ahead of time by the teacher in 85 or higher percent of the cases—and sometimes known by the students as well.

Teachers seldom think of, much less use, instructional strategies that incorporate what we know about learning, especially learning from the constructivist perspective. Nancy and Bob Stahl have provided rich examples of how academic goals can be approached, how instruction can be varied and involving, how academic learning tasks can become active and dynamic, and how success with regard to reaching integrated goals can be achieved. In addition, they advocate an approach that requires students to use systematic decision-making strategies to process and learn concepts and applications.

Decision making is itself a basic skill that everyone needs. It is basic to the comprehension, pursuit, and use of science and technology.

Clear examples of decision-making strategies and the several types of thinking required for teaching and engaging in reasoned, deliberative decision making are provided.

Teachers are given a rationale for using decision-making episodes within cooperative-learning settings. But a rationale is not enough. A total of 33 student classroom episodes (activities) are included as the major component of the book. Ideas have no real meaning unless there are examples and a chance to comprehend these ideas through practice and direct experience. The episodes provide student resources for teachers and serve as a mechanism for teachers to experience constructivist practices in a nonthreatening environment. They provide classroom tasks within which students engage in thinking and learning consistent with constructivist principles. At the same time, they enable students to engage in meaningful interaction with issues, data, and problems important to SST and STS.

The authors have succeeded in illustrating what we know about learning and how teachers can continue to learn from their own students. The mental engagement of students during these 33 episodes encourages them to construct new ideas and to test them in a give-and-take decision-making, social, out-of-the-classroom context. In addition, they reflect upon their existing conceptions, ideas, and databases and make decisions upon the adequacy and uses of these to real-world situations. Such engagement with personal contradictions and applications ensures learning. The resulting student engagement and achievement will stimulate teachers to use more such episodes and to search for other ways to involve their own students actively.

Society and Science offers ideas designed to change teachers, to gain student involvement, and ultimately to stimulate learning for both teachers and learners. The book, which includes philosophy, analysis, applications, and situations ready for immediate classroom use, is for cutting-edge teachers. Use and reflection concerning the content and the activities are sure to exemplify the positive changes advocated by current reformers.

Congratulations to Nancy and Bob for producing an exciting and useful book that pulls together research results, current learning theory, practical classroom learning activities, and reform recommendations in a way that teachers and students can learn by active involvement and direct experience.

Robert E. Yager

President (1992–93)
National Association for Science, Technology, and Society

Professor of Science Education
The University of Iowa
Iowa City, Iowa

PREFACE

During the 1980s, the focus of science education became educating all citizens to participate in an increasingly scientific and technological world. Citizens were urged to become scientifically literate, though definitions and standards for what constituted scientific literacy varied. Many definitions include an understanding of the nature of science, scientific knowledge, and the methodologies of scientific inquiry. *Society and Science* provides *instructional tasks by which students can construct and test ideas and perspectives to enable a strong sense of the meanings and uses of science as it affects and is affected by humans in the modern world.* Many of these ideas and perspectives will challenge students' and teachers' existing beliefs. We expect that students and teachers will accept this challenge with enthusiasm.

Society and Science features a set of 33 learning episodes for middle and secondary school students. The episodes allow students to study and explore many interactions and interrelations among individuals, society, technology, and environments. Students will explore science as *inquiry,* science as a *field of study,* science as a *knowledge base,* science as an *invented way of thinking by humans in the context of their societies,* and science as a *basis for technology*. Students and adults who have used and field tested these episodes have acknowledged that such activities generate new insights about how science and technology affect and are affected by humans as individuals, as members of a society, and as members of a global community.

Every episode has a clear emphasis on humans making decisions that *will* or *may affect* ideas, methods, or uses of science or technology, as well as on how humans *may be affected by* these ideas, methods, and uses. The episodes enable students to study subject-matter content in these three areas, explore interrelationships among them, and practice specific decision strategies. One or more of five specific decision strategies are embedded in each episode.

These episodes include a decision-making emphasis because children and adults alike cannot avoid confronting problem situations or making decisions. In many cases, people face problems without the information or abilities needed to guide systematic decision making. Many do not have the ability to generate viable alternative responses to cope effectively with the situations they encounter. Even more commonly, people are not sufficiently skilled to make competent choices from among the alternatives they do have.

Students should learn systematic ways to deal effectively with situations and alternatives around which a decision must be made. If students are to acquire effective decision-making abilities, they must become actively involved in their own learning. They must engage frequently in worthwhile, process-oriented learning activities that emphasize the specific components of specific, transferable decision strategies. For teachers at all grade levels, helping students acquire effective decision strategies is an important component of classroom instruction. Activities such as those here can help students master valuable decision-making strategies.

The episodes are embedded with large quantities of factual information and subject matter typically stressed in today's classrooms. Students will begin to develop in-depth perspectives on the nature of science and the influences of science on society as well as of society on science. They will study the meanings of science and technology in specific situations. They will also explore the issues of technology—what it is; who invents it and who uses it; what it has, is, and may do; who should control its uses and development.

During these episodes, students become actively involved in making decisions. They respond as participants in scientific, technological, school, and environmental problem-solving situations. In many of these episodes, students are asked to take the role of various major and minor characters in different activities and occupations. In these diverse roles, students confront problem-solving situations from perspectives often quite different from their own. This role-play participation gives students the opportunity to interact with others, to gain information from and about different points of view and perspectives, to clarify and refine their own ideas and beliefs in the context of real problems, and to encounter firsthand a few of the many situations scientists, environmentalists, and people like themselves face today and will continue to face.

You may use these materials to complement or replace existing activities emphasizing the society, science, and technology topics stressed in these episodes. As you use these materials, you will find your students excited by the situations they encounter, the academic content they study, and their personal active involvement in the decision-making phases of each activity. In short, students will be turned on to the subject matter because these activities bring it to life through their active participation in their own learning.

We have seen many positive results from the use of these episodes, results that amazed teachers and turned on students. Such results have always followed a teacher's efforts to use the episodes as part of classroom activities that required pre-lesson preparation, in-lesson monitoring and guidance, and post-lesson follow-up. Teachers have a number of decisions to make prior, during, and after students are engaged in the episodes. The quality of these decisions and subsequent teacher involvement will affect the quality of student involvement and the learning that results. When used appropriately, teachers like what students do during these episodes and learn as a direct result of completing them.

Nancy N. Stahl
Robert J. Stahl

CHAPTER 1

Individuals, Society, Science, and Technology: Ideas and Perspectives

Science *is* and *does* what humans want and shape their inquiry to do. Science does not exist outside the human mind and human decision-based actions. Furthermore, nothing is a part of science until humans decide that it fits one or more conceptions of what science is. For example, the stars are not science, though they can be studied by scientists. Stars also "belong" to astrologers, artists, poets, writers, children, and lovers, among others. Consequently, including a particular objects or content, such as stars, space, or insects, in science textbooks does not make either the object or data "science" or "not science."

Unfortunately, nearly all children and adults carry a conception that particular objects and events in the physical universe, such as animals, plants, rocks, and oceans, are science and belong exclusively to science. This is a misconception. No physical object or phenomenon in the universe—stars, planets, dinosaurs, the human heart, habitats—is exclusively the property or domain of science or scientists. A thing is not within the domain of science until someone decides that it is. This decision and all others pertaining to science, including what science is, was, or will be, can be changed, supported, or refuted by any person. A particular thing becomes part of the domain of science by one's definition of science, by how it is studied, and by the perspective a person uses to investigate and report findings about it.

Therefore, to say that studying the planets and space travel, for example, is to study science both misrepresents science *as inquiry* and *as a perspective* and reinforces the notion that certain things in the world are automatically within the domain of science. Eventually people come to believe that the objects and whatever data are generated about the objects are science. In school settings, and in many beyond-school settings, science becomes whatever data are available and disseminated in "science" books, "science" magazines, and "science" television specials. These conceptions of science must be rejected.

What, Then, *Is* Science?

There are many definitions of science; some are included in the episodes found in this book. Everyone will have their own definitions. Each person, including students, must reflect upon their definitions and assumptions about science to replace misconceptions and to strengthen ideas consistent with contemporary perspectives of what science is.

As students complete the episodes, their notions of the meaning of science will be challenged, including many of their assumptions about and conceptions of scientists, uses of scientific information, and the results of scientific activities. Table 1 contains a partial list of assumptions for use in assessing and making adequate one's concept of science.

Because science is invented by humans, no one owns science—although each person "owns" the conception he or she has of what science is. This self-invented conception is what science is or is not for that person—and will be until he or she modifies that conception. Indeed, science did not exist until humans decided to invent it to help them make sense of the physical world and universe. To change what science is, we merely need to change our definitions of and meanings for it. This is why we can say that what science is to you may be very close to or very different from what science is to someone else. The more different these conceptions are, the more varied will be the perspectives by which people consider and make judgments about society-science-technology issues and relationships.

However, we are not free to invent any definition for science. Notions of what science is are shaped by a person's particular social groups and society at a particular period of time. Ultimately one's definition and conception of what science is must interact with and take into account the perspectives about science that one's society has at the moment and had previously. Consequently, one may adhere to all or part of past or present societal conceptions of science and technology, integrate the two or reject both, or invent entirely new conceptions. In every instance, the conception is invented within the person according to his or her value system.

Science cannot apply itself or make decisions independent of the human mind. Therefore, we cannot claim that science *does* this or that. We must accept that humans choose to use something considered scientific to do this or that. To speak as though science were a person allows us to attack it, praise it, and feel no responsibility for negative consequences of human uses, abuses, misuses, and nonuses of it. For example, to say "Don't blame me, science did it!" shifts responsibility from one's self to science, as though science were an entity that had a brain and could make decisions. While this notion of "science with a brain" is preposterous, people often evoke such notions to remove themselves from accountability for what humans do "in the name of science." Furthermore, *science cannot tell us anything,* but we can use science as a perspective and set of methodologies to describe, make sense of, and explain things.

There is no single scientific method. Inquiry and discovery follow many paths, some less complex, some faster, some more challenging than others—but always there are multiple ways to engage in scientific investigation. Over time, more methods of doing science have emerged, and they will continue to emerge.

Scientific methods are useful for forming new ideas and for testing existing ideas. All too often, students are assigned work focused on testing or "discovering" ideas about phenomenon that are already well established. These ideas—usually presented as facts, concepts, definitions, principles, laws, descriptions, and causal explanations in textbooks and lectures—are provided in ways that suggest that to learn and understand science is to memorize and recall information. In addition, many students

Table 1
A Non-Exhaustive List of Assumptions About and Conceptions of Science

Below are a number of assumptions people have invented about science, scientists, and scientific findings. Read and paraphrase (summarize in your own words) them.

To develop a perspective of science consistent with that of a large number of contemporary scientists across a broad range of fields of inquiry, one should assume that

- Science can only occur within the human brain and "mind"
- Science is a set of perspectives and an orientation toward thinking and the world that are unique and learned—and, hence, can be learned
- Science, as a set of perspectives and orientation toward thinking, cannot exist outside the brain, although information gained from and about this thinking may be recorded outside the brain and studied
- Science is a dynamic, ongoing mental activity that is often accompanied by physical activity
- The world and universe can be comprehended and explained in systematic, meaningful, precise ways
- Scientific inquiry and explanation are often pursued out of curiosity
- Scientific information, findings, and methodologies are the creations of the human mind
- Scientific information and explanations are human-invented tools to describe and organize one's perceptions of the world: they are not true or false but are more or less accurate and viable
- Scientific explanations can often be empirically tested and verified
- Scientific explanations require evidence and consistency with the evidence
- Scientific knowledge is not absolute and is subject to change
- Conclusions and findings from scientific research are always tentative, uncertain, and subject to revision or replacement
- Scientific inquiry cannot solve all problems or answer all questions
- There are many methods of doing scientific research and investigations
- Scientific inquiry and explanation stress comprehensiveness and simplification (rather than complexity)
- Information and ideas generated from scientific thinking are arranged by humans into specific organized fields, areas, and subareas that make sense to humans
- Scientific information, findings, and explanations are not, in themselves, morally good or bad; it is their uses, abuses, and nonuses that may be assessed in terms of morality
- Scientific explanations, conclusions, and principles can often be used to make predictions
- Scientists are obligated to pay attention to, recognize, and avoid using personal and social biases and stereotypes, although some scientists fail to do these things
- The perspectives and orientations of science are not authoritarian, though individual scientists may be
- Scientific thinking always occurs within the context of society and culture and often is itself a social activity

have a misconception that the only "real scientists" are those who discover new things, like finding a new dinosaur's fossils, and write about their discoveries. Science can "exist" and be "done" in the classroom as well as in countless situations outside the classroom.

In every case, whether one is *doing science* rests on critical definitions and assumptions of what science is and requires, and on the methods of inquiry, verification, and explanation used. If one is to think and act like a scientist, one must develop a conception of science that is not bogged down by stereotypes, concrete examples, and faulty reasoning. He or she must come to view science as a dynamic way of thinking that opens up avenues of investigation and topics to study and learn. Scientific thinking at its best is proactive, ongoing, and systematic, and it seeks answers.

A "final" definition of science is not given here, for to do so may lead you to memorize it as so many have memorized *the* steps of *the* scientific method as though there is only one scientific method. Instead, these comments, along with the details in Table 1, can be used to construct a definition and conception of science that will work amazingly well with comprehending science for the modern world.

What Is Technology?

Much of what is mentioned earlier for science also holds for technology. Technology cannot exist without humans who decide to develop, use, misuse, and cease using it. Technology is a great multiplier in that it is *a means* to extend human abilities, to test ideas, and to modify things in the world. Without technology, humans could not fly or sail. Maley (1987) tells us that technology is a mechanical or physical device that, by its application, helps individuals and societies by extending human abilities.

Maley's definition should not be accepted without analysis, for what constitutes technology is a human decision that allows for the many definitions one encounters. According to *Educating Americans for the 21st Century* (National Science Board 1983), technology "consists of the tools, devices, and techniques that have been created to implement ideas borne of science and engineering. Technology exists to manage and modify the physical and biological world in a constructive way." This is one definition that is being used to make sense of what humans have invented and use in the name of "technology." Table 2 lists eight other definitions that may be used to form a more precise definition. Space is provided for you to enter other definitions you want to consider, including your own. The confusion that alternative definitions generate helps to explain why many feel comfortable and secure defining technology, like science, in terms of categories of physical objects (e.g., tools, machines).

Definitions such as those in Table 2 must be considered by each individual in shaping his or her personal conceptions of technology. These conceptions are revised or reinforced via student completion of the episodes in this book as students explore how technology is affected by and affects individuals, groups, and the environment.

The interrelationships and differences between technology and science are not agreed upon, although efforts have been made. For instance, Gies claimed that "Technology is not . . . science. . . . Technology is tools, machines, power, instrumentation, processes, techniques." (Gies 1982, pp. 17–20, in Maley 1987).

The definitions of technology and the assumptions of science may be used to generate descriptions of linkages and interactions between science and technology. The linkages and interactions must be explored actively because the nature of these relationships in the

Table 2
Definitions of Technology

Below are a number of definitions of technology.* Read and paraphrase (reword) them.

1. The means by which humanity extends its potential including tangibles and related adaptive systems.
2. The mental or physical activity by which a person alone, or together with others, deliberately tries to change or manipulate the environment.
3. The efforts by humans to cope with their physical environment—both that provided by nature and that created by their own technological deeds, such as the building of cities—and their attempts to subdue or control that environment by means of their imagination and ingenuity in the use of available resources.
4. A simple tool, a complicated set of machines, a network of overvaulting social relationships, and a general socio-technic for mastering the environment.
5. A disciplined process using resources of materials, energy, and natural phenomenon to achieve human purpose.
6. The cumulative sum of human-made means used to satisfy human needs and desires and to solve specific problems in any given discipline.
7. The technological means undertaken in all cultures that involves the systematic application of organized knowledge and tangibles [tools and materials] for the extension of human faculties that are restricted as a result of the evolutionary process.
8. A body of knowledge and the systematic application of resources to produce outcomes in response to human needs and wants.
9.
10.

* Taken from (1) Karian 1991, p. 6; (2) after Forbes 1968, p. ix; (3) after Kranzberg and Pursell 1967, pp. 4–5; (4) after Hetzler 1969, p. 31; (5) Gradwell 1988, p. 31; (6) after Markert 1989, p. 11; (7) Pytlik, Lauda, and Johnson 1985, p. 7; (8) Savage and Sterry 1991, p. 7.

world are many, often change, and are open to interpretation. Systematic exploration of multiple examples of such relationships and opportunities to make decisions affecting humans within the context of these interactions can enhance one's notions of technology as well as its connections to society.

Addressing Interactions Among Individuals, Society, Science, and Technology Within Curricular and Instructional Settings

Since the first humans began to inquire why things were the way they were and how things could be used or made to adapt to the environment, individuals and social groups have been inventing and using science and technology. At the same time, they have had to respond to, assign meaning to, and appraise the value of both science and technology within the context of their society and environment.

The science and technology that have been invented and used have had positive and negative impacts on society and the physical world. Whether something is positive or negative is determined by the perspectives and values used to assess it. These perspectives and values are personal and social. What one person considers to be positive in a society, others may find to be quite negative (e.g., nuclear power plants). In other instances, what one society may deem beneficial for its purposes, another may find to be negative and destructive (e.g., using science and technology to hunt whales).

These perspectives and values are generated as cultures attempt to deal with the changes and developments in the modern world. Few individuals take time to consider their personal and societal perspectives, as well as the perspectives of others, before taking action that may affect others. Of equal importance, the perspective-inventing and perspective-testing processes do not always consider local, national, regional, and global viewpoints or impacts. Even within the same society or nation, people vary considerably in their perspectives on the uses and nonuses of science and technology. Unless individuals actively engage in frequent and structured interactions with others, they are not likely to fully comprehend their own perspectives, much less comprehend and empathize with the perspectives and values of others. The failure to interact contributes to misconceptions that science and technology have minds of their own, will do what they want to do, and have control over humans—a conception that is dysfunctional, even dangerous, in our complex and interconnected world.

As long as humans exist, the value and uses of science and technology will be continual concerns as people attempt to invent, reflect upon, and use science and technology for personal and social reasons. If their choices and these uses are to be more productive and less destructive, we must stop taking for granted the many positive uses and nonuses of science and technology and critically examine many of the negative. The interactions among humans, science, and technology must be systematically studied lest we continue to pay the extreme human and environmental costs of ignorance, apathy, action, and non-action.

The Need and the Future

Whether we consciously act or not, our day-to-day decisions and behaviors affect science and technology and their development and use. If we are to make effective decisions in a democracy concerning science and technology today and in the future—decisions that will affect each person and the world—we must consider and weigh alternatives. These alternatives will range from definitions and methodologies of science, to what uses will be made of the findings of scientists, to what uses and changes must be made of technology. If one is uneducated in these areas and uninformed about the current status of science and technology, decisions will be made from ignorance, misinformation, or pure self-interest.

The Purposes of This Book

Being concerned about the environment, the uses and abuses of technology, or the way of life of humans on this planet has little meaning if individuals lack the information and abilities needed to shape decisions and take action to address these concerns in positive ways. Part of what is needed is a skillful and intentional move beyond the study of scientific data to the possible, likely, and known applications of scientific and technological information within various social contexts.

By placing students in situations where they must actively consider relevant issues, policies, and details, science and technology can provide the training needed to handle similar situations successfully in their lives outside the classroom. They need to work now so that their future—as well as ours—will be positive, productive, and preserving of human dignity, quality of life, and life and resources on this planet.

The episodes in this book lead students in a search for answers to questions of what, where, and how. The *what* question concerns the particular issues, content, concepts, and methods of inquiry to be studied and abilities to be acquired. The *when* question concerns the appropriate times for scientific inquiry and for the use or nonuse of scientific and technological concepts and abilities. The *how* question focuses on the particular strategies and methods of engaging in scientific thinking and of using science data and technology. These three questions are important because many people already assume that issues and inquiries into social, science, and technology (SST[1]) interactions occur naturally within existing curricular and instructional conditions. The advocacy of SST education, as well as the concern for accurate personal and social conceptions and uses of "science," have emerged because these are not being studied systematically in K–12 classrooms.

A major purpose of this book is to provide students with worthwhile academic tasks that enable them to explore many aspects of science through the eyes of single individuals and groups of individuals. These aspects include the characteristics, methodologies, uses, and results of science. This is a book to seriously consider society and science, because for many students (and adults) the nature, methods, fields, and uses of science are distant, vague, overly technical, and beyond comprehension and control. This is not true in and of itself, but has been true because of the public perception. If we are to comprehend how science can only be an outgrowth of and is always shaped by individual and social values, attitudes, and decisions, then individuals will increasingly feel alienated from and at the mercy of "science"—*as though science existed independently of the human mind and was superior to human thoughts and values.*

A second purpose is to help students practice systematic yet highly transferable decision strategies while exploring and mastering important academic content, concepts, and

[1]We prefer to list these three as society-science-technology (SST) rather than the traditional science-technology-society (STS) because both science and technology are inventions of humans in social groups and their contents, purposes, and uses are all outgrowths of human decisions. In actuality, neither science or technology would exist or would continue to exist without the human decision maker.

principles. In one sense, these episodes reveal that teaching and learning about thinking processes and content can take place simultaneously without reducing either. Teachers can select science topics, issues, and content first and then consider relevant decision strategies their students should practice. In other instances, they may select a decision strategy followed by the important science topics, issues, and content. With sufficient guidance, teachers can learn to construct episodes like these on their own.

Because of the impact of appropriate cooperative-learning tasks, many of the episodes include elements of cooperative learning. Students will have opportunities to cooperatively share their perceptions, make individual decisions, and use negotiation and consensus strategies to arrive at a collective response for which all feel ownership and responsibility. Chapter 4 outlines decisions and activities teachers and students must complete to attain the many personal, social, and academic benefits of students working cooperatively *in groups as groups.*

The book serves a third purpose as teachers attend to the social interaction and participation aspects of decision making and scientific inquiry and allow students to work in cooperative groups. The episodes allow students to work as academic colleagues on academic teams in ways similar to individuals who work in groups in scientific, scholarly, business, and political settings. Students' cooperative civic and social participation can improve humankind. Furthermore, working in groups and as groups enables students to consider a wider range of reasons, consequences, policies, and decisions that may shape, limit, or determine the uses of scientific and technological principles and their affects on individuals, society, and the environment.

The likely abuses and misuses of science and technology by individuals working alone should be enough to want us to start students early in avoiding isolated decision making and in valuing the importance of cooperation and multiple inputs during all stages of the "scientific process" and "technology development and applications."

A fourth purpose is to enable teachers to develop an alternative perspective on student thinking and decision making, a perspective that relies on information-processing models of cognition. Through the models presented here, teachers have an opportunity to generate a view of cognition in which the language students speak or write can be used to infer the kinds of thinking they are engaged in. More importantly, the model for designing episodes like those included here enable students to engage in appropriate content-aligned and decision-strategy-aligned thinking as long as students are on task and completing the activities. As teachers come to comprehend and apply this approach to thinking, they are likely to align students' thinking and learning with what they want students to achieve via their studies.

The fifth purpose is to enable teachers and students alike to construct new, as well as to reflect upon their current, conceptions of science and technology as they are influenced by and influence individuals, groups, and society. Information and ideas related to this purpose were addressed in the beginning of this chapter. Teachers are encouraged to study the ideas presented and to find ways to have their students consider them during their classroom tasks. Having students consider these ideas in combination with the activities in this book will make a strong impact on their perspectives of science and technology and on human beings as decision makers, as creators and users of science and technology, and as beings that affect and are affected by the multiple conceptions and uses of science and technology (also see Casteel and Yager 1966).

Ultimately the episodes and the contents of this chapter will help students and adults alike rethink their conceptions and assumptions about science and technology; their meanings, uses, and nonuses; and the reasons for and consequences of their uses or nonuses. Our intention is to bring about worthwhile examination of the individual, as an individual in societal contexts and as a decision maker, on the shaping, assessing, and use of science and technology. This examination will most likely lead to decisions and actions that show greater attention to the diversity within science and technology as well as a deeper concern for individual humans and for the environment that is our mutual spaceship in this universe.

CHAPTER **2**

Deliberative Decision Making: A Framework for Instructional Tasks

All too frequently decision making is equated with problem solving. The two processes are not the same. When you take steps to solve problems, you do more than merely make a decision. Selecting what to wear or what to eat are examples of decisions you make daily. Such decisions don't usually become problem-solving situations. In most instances, you encounter situations where decisions are needed or are made in the absence of problems. Problem solving, on the other hand, is geared to making decisions that are designed to resolve or significantly enhance the handling of a particular problem, dilemma, or conflict. Hence, while problem solving requires that decisions be made, decisions can be made in the absence of a problem situation.

Whether problem solving or not, *deliberative decision makers* tend to make more adequate decisions after their reasoned consideration of the situation, available information, possible options, and possible implications. The more important the decision—especially one embedded within a problem-solving situation—the more these decision makers tend to consider relevant details other than just what decision to make.

In situations requiring important decisions, deliberative decision makers systematically and continually take time to think through relevant information. This frequently involves locating and considering relevant details, defining terms, generating lists of feasible options, selecting standards and reasons for making the most appropriate decision, and exploring who and what may be affected by the decision. These decision makers take a serious look at the long- and short-range consequences of their options prior to their decisions. In the midst of their deliberations, they empathize with others in the situation and with those who may be affected by the consequences of the decision. They know that making the decision is only one of a series of interrelated steps that together move them toward the best decision. They accept the notion that the final decision usually implies that action will be taken. In other words, they view the decision itself as a descriptor of the action one or more persons should, must, or will carry out.

On the other hand, *nondeliberative* (or hasty) *decision makers* focus on the immediate making of a choice. They fail to spend sufficient time generating and studying possible options. They typically ignore or discount long-range and even short-range consequences and the extent to which the decision and subsequent actions may affect themselves, others, and other things. They tend to avoid (a) considering or seeking out new information; (b) reflecting upon data contrary to their present perceptions; (c) spending time exploring alternative interpretations and conceptions; and (d) considering the appropriate standards for making the most appropriate decision. For these reasons, nondeliberative decision makers frequently (a) re-encounter similar situations; (b) take no or little responsibility for their own decisions; (c) are less committed to adhering to their decisions via subsequent action; (d) are unprepared for the consequences of their decisions; (e) blame fate, chance, or others for unfavorable results; and (f) have greater levels of negative emotions for longer periods of time after they make their decisions than their more skilled, deliberative counterparts. In their haste and in their use of inappropriate strategies, they make ineffective and unsatisfying decisions. Consequently, they find themselves dealing with unanticipated results that in turn lead to greater and prolonged feelings of uncertainty, frustration, anxiety, and even panic.

Because deliberative decision makers tend to consciously consider prior to the final decision the context and the situation, the available information, the relationship of the situation to other people and things, and the implications of their decisions, they prepare themselves to live with many of the consequences of their decisions. They feel relatively satisfied about the decisions they make—even when the setting requires restrictive or unpleasant options.

Although prior considerations can never eliminate all negative consequences and feelings, the deliberative decision maker values conscious, reflective predecision considerations because these efforts greatly increase the probability of positive feelings and consequences they can live with in the future.

Deliberative decision making is a learned way of approaching the making of decisions. *Because it is learned, it can be taught and mastered in school.* Students can learn to think and decide deliberatively when they have appropriate procedural information to guide their thinking. Step-by-step procedures aligned with one of five decision strategies are included in each activity in this book. The next section provides details concerning the five decision strategies that are embedded in the structured learning episodes in this book.

FIVE DECISION STRATEGIES

Five distinct decision strategies students can learn to use are integrated into the structured learning episodes. They are not the only decision strategies students could or should use, but they are ones many people frequently use—or think they use—when they attempt to make the most appropriate decision in everyday situations. There is no hierarchy to the five strategies; they are simply five alternative approaches to making decisions and solving problems.

These five decision strategies are described below. Use these descriptions to help students stay on task as they complete the tasks required in each episode.

The Forced-Choice Decision Strategy

This strategy is useful when students have a situation, problem, or dilemma in which the options given are

- limited in number
- all specifically stated

- homogeneous; for example, they are either all "good" or all "undesirable"
- clearly stated or recognized within the situation and context
- the only ones that can be considered

With the forced-choice decision strategy, students are required to make a choice from a limited number (usually three to six) of almost equally desirable or undesirable alternatives.

Students must come to accept that to refuse to make a decision takes the matter out of their hands and almost always increases the risk of a worse consequence than any of the options already available. In forced-choice situations, the student must focus on selecting only one option from those available. Forced-choice situations do not allow students the luxury of inventing other options or combining given options; to spend time in trying to do so actually reduces the time they can spend on the options that can be selected. Once the decision is made, the nonselected alternative options are no longer available.

Seven of the episodes in this book require the forced-choice decision strategy.

The Rank-Order Decision Strategy

This strategy is useful when students have a situation, problem, or dilemma in which the options available are

- limited in number
- all specifically stated
- homogeneous, such as either all "good" or all "undesirable"
- clearly stated or recognized within the situation and context
- the only ones that can be considered
- selected in terms of the priorities of the student or group such that they are arranged in order from most preferred to least preferred

This decision strategy is useful when students must make a decision based upon the level of priority assigned each alternative.

Students must accept that the only options available are those provided and that each option is independent of the others. In many of these situations, the student must accept that his or her first-ranked choice may be unavailable or may not work to resolve the problem. If the first choice is not available or does not work as expected, the second-ranked choice will be made available. Students follow this priority procedure until all available options have been ranked. In other words, students rate options in terms of their priority, importance, value, or usefulness to them in that situation at that moment.

Typically when people are asked to rank-order options, they operate as though their top-ranked choice will always work; hence they pay little attention to their remaining rankings. The perspective for rank ordering options used in this book reminds students that they must continually select the highest-priority option from the remaining alternatives, down to the last two options. This strategy forces students to constantly consider the relative importance of all alternatives within the context of a particular situation.

Six of the episodes in this book require the rank-order decision strategy.

The Negotiation Decision Strategy

This decision strategy is useful when students must make a decision in which the options given are

- limited in number
- all specifically stated
- homogeneous, such as either all "good" or all "undesirable"
- clearly stated or recognized within the situation and context

- the only ones that can be considered
- they must agree to give up certain options to gain others they want or desire the most

In these situations, the student or group must divide the alternatives into three specific classes of options (such as alternatives that are most desirable, those they are most willing to give up, and those not included in either of these two groups). They place options into the classes based upon those they most want, those they are willing to give up to get what they most want, and those that are neither the most desired nor the least acceptable.

Using the negotiation decision strategy, students operate in a compromise mode in which they must decide what they most want and what they are most willing to give up. They must systematically negotiate and compromise with themselves. They are forced to consider the relative importance of other alternatives in situations where they have to give up something they may need to gain, preserve, or protect other things they see as being even more important. Students often need to employ this decision strategy in many everyday situations. In small groups, they must practice negotiation and compromise abilities with others.

Five of the episodes in this book require the negotiation decision strategy.

The Invention Decision Strategy

This decision strategy is useful when students are to make a decision in which they are relatively "free" to make any decision consistent with and appropriate for the particular situation or problem

In some of these situations, students may have a set of possible options. They may select one or more of the options, reject them all, or combine some of them to form new options. In all cases, students are free to invent or generate any decision that fits the context of the situation—but they *are not free* to ignore or reject the context and the likely consequences of the decision.

The focus of the invention decision strategy is for students to construct the most appropriate decision to respond to and resolve the situation given constraints and limitations. For this reason, the invention strategy is also called the *open-ended* or *free-response strategy.*

Six of the episodes in this book require the invention decision strategy.

The Exploration Decision Strategy

This strategy is useful when students encounter written or graphic information or a situation they want to examine in greater depth to form a reasoned response.

In these situations, students need to ask appropriate questions to focus their comprehension of the content, context, and situation; their decisions about the relevancy of this information, situation, and context to one or more issues, concepts, events, or decisions; the alternatives, consequences, and criteria they should consider relative to the issue; and their preferences, emotions, and values as they perceive them to be relevant.

This decision strategy includes questions students ask and answer for themselves about what is being studied or learned. The questions represent at least four broad types of thinking—conceptual, relational, decisional, and affective (described with examples in the following section)—that are transferable to a wide range of situations and data. To guide their own learning, students should master the ability to generate their own questions relative to these four general types of thinking. While the episodes here include sets of questions, you should see these as examples of the kinds of questions students should learn to ask on their own.

Nine of the episodes in this book require the exploration decision strategy.

FIVE TYPES OF THINKING RELATIVE TO MAKING DECISIONS

Five generic types of thinking skilled decision makers use as they move toward their decisions are described below.

Conceptual Thinking

This type of thinking stresses adequate comprehension, description, and clarification of available information; of a situation, phenomena, or problem; and of the meaning of data and relevant concepts. Conceptual thinking includes generating meanings and interpretations of the situation and data; clarifying terms; determining the major point or focus of a situation, set of data, or problem; and assembling relevant descriptive information about persons, places, things, and time involved. In addition, conceptual thinking involves inventing such information and conceptions as appropriate definitions, attributes of concepts, meanings, and interpretations as the student attempts to make sense of the things being considered. Conceptual thinking answers such questions as

- What details about this event are known?
- What does this mean?
- What is your interpretation of this situation? These data? This action? This chart?
- What definition are you using for this term?
- What are the relevant attributes, characteristics, or features?
- At what location did it occur?
- At what time did it occur?
- What description do you have of it?

This type of thinking involves the comprehension, sense making, and description of the problem; and of relevant information, events, people, details, and contextual information surrounding the problem.

Relational Thinking

Sometimes referred to as "connective thinking" or "thinking for relevancy," relational thinking emphasizes the interrelatedness of and associations among information, a situation, data, or context to a student's or group's past or present databases, abilities, beliefs, meanings, and needs. Relational thinking stresses the student's acknowledging and constructing linkages and associations among information, situations, verifiable details and data, concepts, beliefs, feelings, and values to previous or present ideas, data, actions, or situations. These associations may also involve linking events, people, things, or data in the academic situation being studied to past, present, or future life situations or needs. Relational thinking answers such questions as

- How are these two facts *connected?*
- How are these data *tied* to the major focus of your research?
- In what ways are these data *linked to* the problem you are trying to solve?
- What is the *connection between* this alternative and your long-range goals?
- How do these data *fit with* the data you collected earlier on this same subject?
- What has this information *to do with* the topic you studied yesterday?

Students would also use relational thinking to connect what they are currently studying to concepts, principles, and data they have previously studied or may encounter in the future. When students describe or invent these interconnections, they determine the relevancy of the information or situation to other things—including to themselves and to what they value, want, need, or know.

Relational thinking involves connecting or associating information, interpretations, alternative solutions, consequences, beliefs, and the

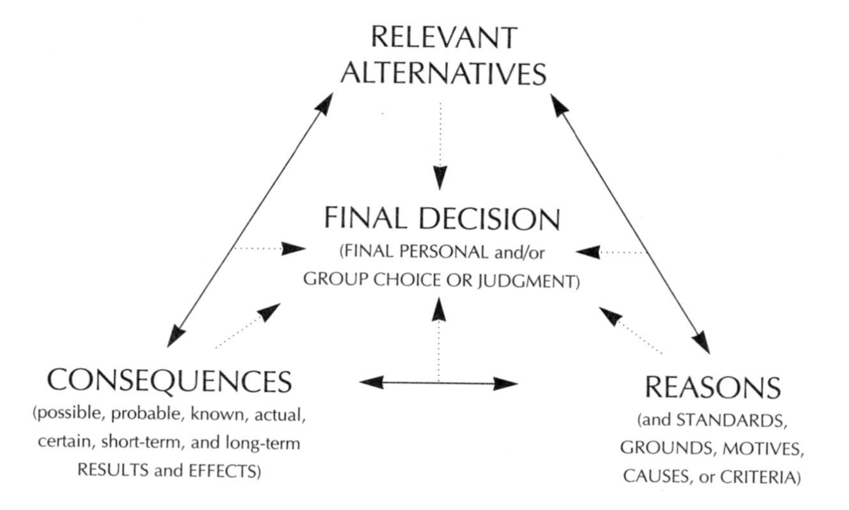

Figure 1

The interaction among essential elements of decisional thinking. This interaction and interdependence of elements are critical to every comprehensive decision, problem-solving, and inquiry strategy.

problem itself, to involved persons and things, and to the student's or group's needs or concerns. Failure to consider and formulate such connections often results in decisions that have little relevancy to the problem or student.

Decisional Thinking

With decisional thinking, students consider, determine, and use specific categories of information as they seek to consider options before making decisions. These categories include consideration of (a) given, likely, and possible choices, actions, or policies; (b) known, possible, or probable consequences of various alternatives; (c) reasons, grounds, standards, or criteria that might be or are to be used as the basis for making a decision; and (d) the advantages and disadvantages of possible decisions in terms of the short- and long-term implications and effects of each option or policy. Figure 1 illustrates the preferred and most important interrelationships among these four categories.

If a student or group fails to deliberately consider available and alternative choices; possible, certain, or known consequences; likely or required criteria or standards; and their own preferences, needs, and expectations in relation to each other prior to a decision, the student or group was not engaged in a deliberative decision strategy. Decisional thinking answers such questions as

- What options are available?
- What standards or reasons are being considered? Should be considered?
- What criteria did you consider?
- What are the likely short- and long-term consequences of this option?
- How might each option affect the people involved?
- What reasons would you give to justify that decision?
- What would be the most immediate results of this decision?
- What consequences do you want to avoid?

Affective Thinking

Affective thinking stresses adequate consideration of personal or group (a) preferences, likes and dislikes, and priorities; (b) emotions and feelings; (c) commitments to decisions and actions; (d) desires and dispositions; and (e) value orientations. As they verbalize these considerations, students are better able to consider how their emotions, desires, and the like may affect or have affected their decisions and actions. Affective thinking answers questions such as

- Of these options, which one do you prefer?
- What is the worst decision you could make?
- What word best describes the emotions you felt at the time you made your decision?
- How committed are you to carrying out the decision you just made?
- What emotions are the people who might be affected likely to have to your decision?
- Prior to making your judgment, what social values did you consider?
- Now that you have made your decision, do you feel relaxed, afraid, or anxious?

As students explore the nature of science and technology within societies and the ways these affect and are affected by individuals and society, they need to objectify and examine their own preferences and values toward science, technology, and the uses and nonuses of both at particular moments. They also may examine their reactions to science and technology in their various forms; to situations, events, and actions involving science, scientists, and technology; and to possible applications of science and technology in their own lives. In addition, students need to systematically consider the preferences, values, priorities, and feelings of others involved or likely to be affected. Affective thinking concerns a personal dimension not disclosed in the first three types of thinking.

Relationships Among These Four Types of Thinking

Table 3 summarizes characteristics of the four types of thinking described above. Deliberative decision makers integrate and interrelate the information in these types of thinking as they move toward making the most appropriate decision. Figure 2 (page 18) illustrates these four types of thinking as they are most likely to take place within skilled decision makers. Figure 2 also shows the thinking patterns students should use as they respond to the episodes.

Reflective Thinking

This fifth type of thinking, reflective thinking, is unique in several ways. It is described separately because students can only engage in it after one or more decisions involving the first four types of thinking have been made. Reflective thinking occurs only after a decision has been made, as students systematically examine the thinking that led to their decision and actions. Systematic reflection involves the ways students processed, assessed, and made use of the available and invented information, conceptions, associations, options, values, criteria, and consequences to reach their decisions.

This after-the-decision deliberative examination should include a careful consideration of the step-by-step procedures the student or group used to reach the decision. To pull together and assess the entire process of decision making, the student reflects upon the data that were considered, how things were considered, and in what sequence and manner they

Table 3
Types of Thinking Engaged in During the Episodes

Type of Thinking	Characteristics
Conceptual Thinking *Comprehending and Understanding*	• state focus of activity or problem • recollect relevant data and information • interpret data • explain meanings assigned to data • state available data • define terms • paraphrase prior statements
Relational Thinking *Identifying and/or Forming Relevant Associations, Comparisons, and Connections*	• state how things are connected • explain associations among things • state how data are relevant • state relationships • state similarities and differences • make comparisons among data and events
Decisional Thinking *Considering Decision Factors and/or Resolving Problems*	• state alternatives and options • state known and possible consequences • consider effects and results • state criteria and reasons • declare imperatives • state a decision by a person or group
Affective Thinking *Considering and Making Value Choices and Expressing Emotions and Feelings*	• state one's personal preferences • cite value ratings and rankings • state value/moral choices or judgments • express emotional responses or reactions • express personal feelings • cite emotions and feelings of others • state personal dispositions or wants

INTERACTIONS AMONG THE FOUR GENERAL TYPES OF THINKING

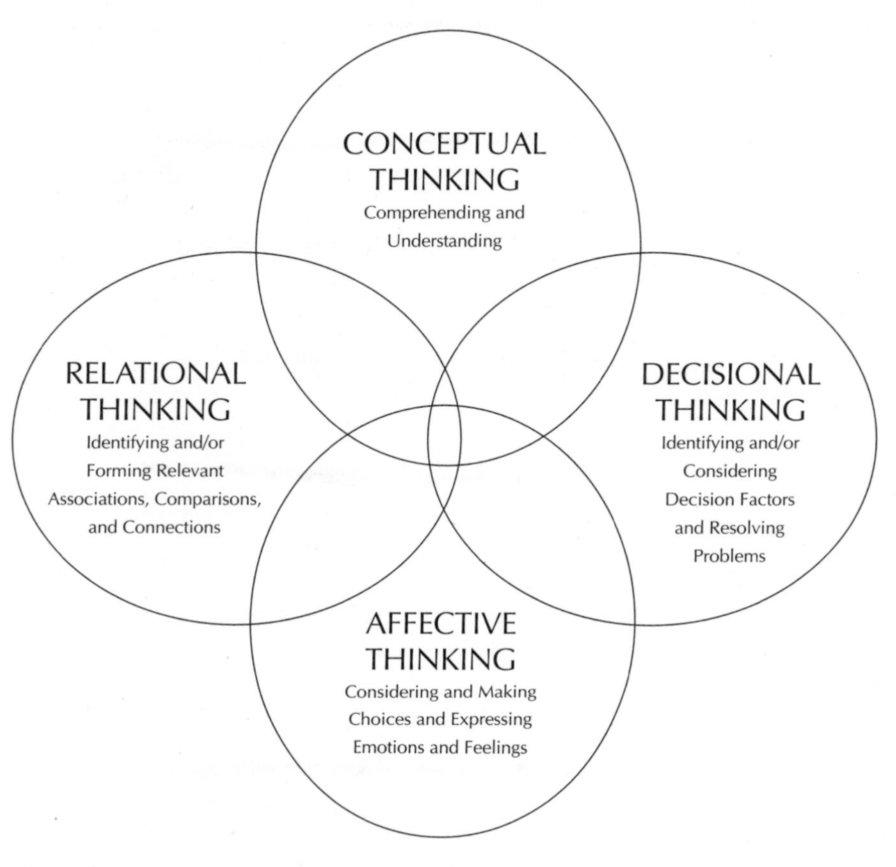

Figure 2

To be most effective, these four types of thinking should take place interactively to complement one another during meaningful learning and decision-making tasks.

were considered. When done adequately, this reflection avoids a quick reassessment of just the decision itself. Consolidation via reflection thinking answers questions such as

- In these situations, how did you determine what the problems were?
- As you tried to make your decision, what steps did you follow?
- To what extent did you use the same criteria across these instances?
- To what extent did you consider the same consequences across these situations?
- What consequences did you continually try to avoid? Try to achieve?
- What standards did you continually try to uphold? Try to ignore or reject?
- To what extent did you use the same definitions across these situations?
- If you had to make this decision over again, what would you do now that you failed to do earlier?

If students are to become skilled at using the five decision strategies, they must use each of these five types of thinking. They must have extensive practice reviewing, reflecting upon, and assessing each step in each of the strategies. To assist in this reflective process, have students answer questions such as those in Table 4 following each episode. Students should also answer them following every activity in which they make important decisions—especially decisions involving their efforts to resolve conflicts and solve problems. (More details on these questions and their roles are provided in the Follow-Up Discussion and Questions section in Chapter 3.)

Such generic questions focus and guide reflection of processes and strategies students followed in making their decisions. More importantly, they assist students in consciously studying their own thinking processes so they may become deliberative decision-strategy users. While you would not ask every question listed, a selection of these questions need to be answered to help students consolidate, reflect upon, and assess their own decisions, decision-making processes, and affective responses.

Teachers typically copy this page and place a check mark on the line next to the specific questions they will ask after a particular episode.

Reflection thinking should be engaged in after one or more learning episodes have been completed. This "reflective phase" of thinking and learning is most effective after a student has studied and made decisions regarding three or more situations relative to the same issue. In this way students can discover the extent to which they were consistent in using the same criteria, concept definitions, and reasons as they sought to make appropriate decisions across different problems and situations. They can consider the reasons for their consistency or inconsistency. For example, students might use the same word, such as *science,* in three different situations, yet use a different definition in each situation. This inconsistency would suggest that the meaning of the word changes from situation to situation. Students need to explore the implications of changing their meanings as well as of keeping one meaning across situations.

In another unit, students might switch the standards they use from situation to situation, which may or may not be appropriate. The point is that students need to systematically reflect upon both their prior decisions and the steps they took to reach those decisions. The fifth type of thinking is one tool you can use to help students engage in this reconsideration and reflection.

* * *

How these five decision strategies are blended into academic learning will become clear as students work through the 33 episodes. Your job is to monitor how well students are adhering to the strategies as they complete their several decision-making tasks. The decision strategies need to be completed in conjunction with the five types of thinking. When used together, correctly, and often, students will acquire five powerful approaches to making decisions.

Table 4

Questions to Guide Student Reflection of Their Thinking and Decision-Making Processes During a Decision-Making Episode

____ **1.** As an individual trying to make a decision in this situation, what was the major problem you faced?

____ **2.** What does it mean to "reach a consensus"?

____ **3.** To reach your final decision, what specific steps did you take?

____ **4.** To reach its final decision, what specific steps did your group take?

____ **5.** As you tried to make your final choice, what alternatives did you consider?

____ **6.** How much time did you spend considering the consequences of each alternative?

____ **7.** What were the best reasons you considered for making the decision you made?

____ **8.** In your group, to what extent were you able to express your thoughts and feelings?

____ **9.** In what ways would your final decision actually resolve the problem you faced?

____ **10.** If you had not used this decision strategy in this situation, what decision strategy would you have used?

____ **11.** When others disagreed with your ideas or decisions, how did you feel?

____ **12.** This episode asked you to choose from a short list of choices. What are two situations in your life where you are limited in the options you have?

____ **13.** When you found you could only choose from among the choices listed, to what extent were you restricted from making the best decision possible?

____ **14.** In most situations when a person needs to make an important decision, things such as time, money, and abilities limit the choices. In this situation, what things limited you in making a better decision?

____ **15.** If you could have made *any* decision, what would you have decided to do?

____ **16.** To what extent were you satisfied with your (or your group's) final choice?

____ **17.** In this situation, to what extent did making your decision also solve the problem?

____ **18.** In shaping your final decision, what facts did you consider to be the most important?

____ **19.** In the situation, what was the "real" problem you were attempting to solve?

____ **20.** In your daily life, how might you use this decision strategy to resolve problems?

____ **21.** In your daily life, what are at least two examples of situations where this decision strategy could be useful?

____ **22.** What are at least two advantages (or disadvantages) of using this decision strategy?

____ **23.** In daily life, what are at least two restrictions on using this decision strategy?

CHAPTER 3

Structured Decision-Making Episodes: An Overview

If you value student comprehension and use of social science, science, and technology content and abilities, including methods of inquiry, you will be excited by these student episodes. Each of the 33 episodes contains society-, science-, and technology-related information. With these episodes as learning tasks, you avoid the difficult choice between students' adequate comprehension and use of essential academic content and concepts or their acquisition and mastery of essential, highly transferable decision-making strategies. Through completion of these episodes, these valued abilities are integrated and attained simultaneously.

The episodes were designed to facilitate students' use of conceptual, relational, decisional, and affective thinking. As students and groups follow the activity directions, they will automatically use the four types of thinking described in Chapter 2. They will also complete their thinking and decision making without a great deal of direct teacher involvement. The episodes incorporate the five distinct problem-solving strategies.

Most of the episodes are designed for students to respond first as individuals and then as members of a group. Integrated in the episode short story are directions to guide individual and then group decision episodes. Students read what their roles are, the steps they follow in considering the problem and devising a solution, and how they move from their individual decisions to reach a group decision. Students are fully involved in their own learning and using the decision-making strategy built into each activity, so you have time to give individual attention when needed. You will spend your energies monitoring students while they consider the situation and relevant details; clarify and reason through their understandings; and reach their decisions.

The first 25 episodes are designed for students to work in groups of two, three, or four. Place students in heterogeneous groups *before* passing out copies of the episodes. Remind students to make individual decisions before trying to reach a consensus group decision. The group consensus discussions can continue until all groups have made a decision. Then each group can report its findings to the class. Or the group discussions can end prematurely so the class can consider the important to-be-learned information processed during the individual tasks and whole-group discussion. The group decision is not important in and of itself: the task of reaching a group decision is a device to get students to review the subject matter and their comprehension of the situation they are facing. The structured interaction with peers enables students to spend time reviewing the details of the situation, their comprehension of those details, the connections they invent or recognize, the reasons they use, and the decisions they make.

Once students have engaged in sufficient interaction toward arriving at their group decisions, use the follow-up questions included with each episode to review the major points and information students were to learn, their decisions, and their thinking as they worked toward a decision.

Not all of these episodes involve problems or dilemmas. In some cases they place students in inquiry situations in which they answer a series of questions in response to a particular issue, object, reading, or course of action. These response activities help students realize how they may structure their thinking in situations where no structure decision strategy is required.

The teacher's role for these exploration episodes is to use questions to guide students' thinking. Questions are used to direct students to consider specific kinds of information as they react to a set of data, a problem, or a situation. In each exploration episode, a set of five to ten questions has been preselected for students to answer in written form to focus them on important information to consider before they begin a group discussion. In addition to these five to ten questions, an extensive list of other questions is provided to further guide the discussion.

FIVE EPISODE FORMATS

The five distinct decision-making strategies described in Chapter 2 have been integrated within the 33 episodes to ensure students use at least one strategy to complete each episode. Consequently, each learning episode integrates at least one of the decision-making strategies.

The word *format* describes how an episode is designed to ensure students use a particular strategy to complete the learning task. The decision strategy embedded in an episode is not an option; it is a required part of the tasks students complete within that activity. The descriptions of the five formats of structured episodes below describe how each decision strategy is incorporated into the episodes that include that strategy.

Forced-Choice Format

Forced-choice episodes give students a problem, dilemma, or situation in which a major character or group has to make a decision. In this situation the options available to the character or group are

- all specifically stated
- limited in number and limited to only those specified
- either all equally "desirable" or equally "undesirable" options
- all stated within the written materials and context provided
- the only ones that can be considered for and included in the final decision

In these episodes, the characters or groups are placed in situations in which they must make choices from a limited number of stated alternatives (usually three to six alternatives are given). The forced-choice format emphasizes situations in which a character or group must decide among a limited number of almost equally desirable or undesirable nonchangeable alternatives. In each episode, it is explained why a character's or group's refusal to make a decision takes the matter out of his or her hands and increases the risk of a worse consequence than any of the options already available. At the same time, the story part of the episode makes it clear that considering other options is unacceptable and a waste of time. The character or group can choose only one option and, once selected, the remaining options are no longer available.

Learning tasks emphasizing the use of this strategy allow students to practice making decisions under forced-choice circumstances. These learning tasks also help students understand the factors that need to be considered in such restrictive situations so a decision can be "lived with."

The episodes illustrating the forced-choice format and incorporating the forced-choice decision strategy are (the contents of the episode is given in parentheses):

Episode 1. *My Favorite Monster* (Herbivorous dinosaurs)
Episode 2. *Remains* (Ownership of Native American artifacts in museums)
Episode 3. *Choose or Lose* (Community conservation projects)
Episode 4. *Too Few of a Good Thing* (Economic concepts applied to ticket distribution)
Episode 5. *The Circle* (Research on death and dying)
Episode 6. *No Deep Breaths* (Air pollution and the health of employees)
Episode 7. *Cliff Hanger* (Hiking and rappelling safety precautions)

Rank-Order Format

These episodes provide students with a story in which a major character or group must make a decision. Here the alternatives available are listed. In this situation the options available to the character or group are:

- all specifically stated
- limited in number and limited to only those specified
- either all equally "desirable" or equally "undesirable" options
- all stated within the written materials and context provided
- the only ones that can be considered for and included in the final decision

In the episodes students are told the only options available are those provided and that each option is independent of the others. These episodes require students to constantly consider the relative importance of a number of alternatives that are nearly identical to one another. They assign a specific priority number (for example, 1 to indicate highest priority, 2 to indicate second highest priority, and so on) to each option.

In these short-story episodes, the characters are deliberately made aware that their highest-ranked choice *may not be available* at the time it is needed or may not work to resolve the problem. Thus, the second-ranked choice will be considered for its availability or effectiveness. This procedure will be followed for all available options.

The fact that students are made aware that their character's or group's choices will be considered by others in the order of their own rank is critically important. In many other decision-making activities when students think their top-ranked choice will always work, they pay little attention to ranking the remaining options. The procedure required to complete these episodes ensures that students will select the top option from the remaining alternatives all the way down to the final pair of options to be rank ordered.

Episodes illustrating the rank-order format and incorporating the rank-order decision strategy are:

Episode 8. *Heavenly Bodies* (Objects in the solar system)
Episode 9. *Oops! The One Time I Forgot!* (Recording scientific observations and experiments)
Episode 10. *Becoming a Master Scientist* (Learning to become a scientist)
Episode 11. *Circle of Poison* (Research on possible poisons in foods)
Episode 12. *Those Pesky Pesticides* (Destructive pesticides and possible effects on business)
Episode 13. *Representing the People* (Environmental balance versus economic prosperity)

Negotiation Format

These episodes provide students with a story in which the major character or group must make a decision. Here, the only alternatives available are listed. In this situation the options available to the character or group are:

- all specifically stated
- limited in number and limited to only those specified
- either all equally "desirable" or equally "undesirable" options
- all stated within the written materials and context provided
- the only ones that can be considered for and included in the final decision

However, in this activity the character or group must divide the available alternatives into three specific groups or classes of options (those alternatives that are most desirable; those they are most willing to give up; and those not included in either of these two groups). The characters in the story must compromise or bargain among themselves to determine which options they most want or need and which they are most willing to give up to secure their most important alternatives. In each negotiation episode, students are told how to sort their options into the three classes or groupings.

The negotiation format requires students to consider the relative importance of all alternatives in situations in which they have to give up something to gain, to preserve, or to protect other things that they perceive as more important. Consequently, students develop the abilities to compromise, to bargain, and to negotiate—abilities they can use in many everyday situations.

Episodes illustrating the negotiation format and incorporating the negotiation decision strategy are:

Episode 14. *Off We Go* (Manned and unmanned space exploration)
Episode 15. *Why? O Why?? O Why???* (Theories on the extinction of the dinosaurs)
Episode 16. *Tough Decisions* (Equipment for science labs)
Episode 17. *We're Running Out of Juice* (Energy conservation in a school setting)
Episode 18. *Scientists—They're Everywhere!* (Occupations in science)

Invention Format

These episodes provide students with a story in which the major character or group must make a decision. The individual or group in the episode is free to make any decision consistent with the given situation. In some episodes that fit this format, they are provided some possible options. They may select any of them, they may combine them to form choices they like better, or they may choose to reject them all.

In all cases, the story characters are free to "create" any appropriate decision they want that fits the context of the situation. During these learning tasks, students are encouraged to select or invent an appropriate decision to respond to and resolve the situation before them.

This episode format helps prepare students for decision-making situations in which they have a great deal of flexibility in what decision(s) they can make.

Episodes illustrating the invention format and incorporating the invention decision strategy are:

Episode 19. *Bo or Zo?* (Subfields of botany and zoology)
Episode 20. *Pulling Teeth* (Scientific issues concerning mercury in teeth fillings)
Episode 21. *A Harey Situation* (Relationship among people, animals, and the environment)
Episode 22. *Whose Fault?* (Possible consequences of construction near a fault line)
Episode 23. *Flighty Decisions* (Possible effects of supersonic passenger planes)
Episode 24. *Hear Ye! Hear Ye!* (Noise, music, and noise pollution)

Exploration Format

These tasks provide students with written or graphic information and a set of five to ten questions to answer in writing. The information or graphic provides a limited set of subject matter for students to consider. The questions help students consider the important information they are to examine, clarify, and use. The questions also help the teacher to monitor student subject matter comprehension, how students relate the subject matter to the focus of the inquiry or learning, and what decisions and judgments students have made about the information and issue presented. These questions should serve as springboards to expand the oral discussion following each student's written answers. In these episodes, the questions are included on the same pages as the original information.

Episodes illustrating the exploration format and incorporating the exploration decision strategy are:

Episode 25. *The Case of the Missing Statue* (One method of scientific investigation)
Episode 26. *A Bird Story* (Prehistoric origins and ancestors of today's birds)
Episode 27. *Iceberg!* (Research on icebergs)
Episode 28. *Athletes' Feats* (Physiology, health, and athletic performance)
Episode 29. *Breakthroughs* (Technology and advances in aircraft design and performance)
Episode 30. *Ye May Keep Your Pet Rocks* (Restricting the importation of injurious wildlife)
Episode 31. *Out of Frustration* (Definitions, meanings, and views of science)
Episode 32. *Dirty Snowballs* (Theories of origins and composition of comets)
Episode 33. *Rain or Shine?* (Analyzing national weather maps

ELEMENTS OF THE EPISODES

All of the episodes in this book are designed to be flexible. Teachers can adapt and implement them according to their students' needs. The Appendix provides a sample lesson plan for several selected episodes. These plans are illustrative and by no means suggest the only way the episodes can be used. The vast majority of teachers who have field tested these episodes developed their own lesson plans for each activity they used.

The Structure of the Decision-Making Episodes

The 33 episodes in this book present structured learning tasks to which students respond. Instructions on how to proceed through the learning task are built into each episode. The following are the typical parts to the episodes for the forced-choice, rank-order, negotiation, and invention formats.

A story or set of information. The story to which students must respond with an appropriate decision places students as characters in a problem-solving situation where they must make an important decision. After reading and studying the story, students write their responses on Predecision Task Sheets, Individual Decision Sheets, and/or Group Decision Sheets.

One or more Predecision Task Sheets. These worksheets, a required part of each episode, serve to maintain the society- and science-oriented content focus by helping students systematically comprehend and consider relevant information. Before they make their final decision, students are encouraged to complete the Predecision Task Sheet(s) on their own and then share their responses in a small group. Predecision tasks help students as characters in the story to systematically and deliberately consider important information relative to the content and context in which they find themselves. They also serve as rehearsal tasks for students to rethink the subject matter they are to learn.

An Individual Decision Sheet. This sheet is where students as characters in the story write out their decisions about the situation and information they have studied. Each Decision Sheet fits one decision-making strategy and helps systematize the steps students should follow for the particular strategy. As often as possible, students are encouraged to work through and answer each Individual Decision Sheet on their own *before* moving on to complete the Group Decision Sheet as members of a small group.

A Group Decision Sheet. This sheet is where students, as members of small groups, write out their joint decisions about the situation and information they have studied. Each Decision Sheet fits the identical decision-making strategy as used for their individual decision. This sheet helps systematize the steps students, as members of a group, should follow for that strategy. When students arrive at a group decision, they use persuasive consensus skills rather than the principles of majority rule.

You may want to distribute the Individual Decision Sheet only after students have had sufficient time to answer the questions on the Predecision Task Sheets. Often when these two sheets are passed out together, a number of students will not complete the Predecision Task Sheets and will hurry on to the Individual Decision Sheet. One copy of the Group Decision Sheet is usually given to each group. When students work in groups, you may want to wait and pass out the Group Decision Sheet only *after* students have had time to complete the Individual Decision Sheet.

Some episodes include Decision Sheets marked as optional. Assign these sheets based on available time and the ability of your students to comprehend the information given in the episode.

Follow-up Discussion and Questions

Each episode should end with a follow-up discussion. You are responsible for guiding this discussion with questions. The questions provided with each episode serve as suggestions. Choose those appropriate for your students and the focus you want to emphasize. You are encouraged to devise other questions to guide student learning appropriate to the episode and academic outcomes.

Review and Reflection Questions: Emphasis on Academic Content. These questions, appearing at the end of each episode, are suggested to help review and consolidate students' thinking and learning relevant to the episode as well as to stimulate further discussion on the content and purpose of the episode. Some questions emphasize the comprehension and clarification of social-, scientific-, or technology-related information students are expected to learn from the episode. Others stress the connecting of science-based ideas, concepts, and information examined in the episode to social or technological information, issues, or actions. Still other questions focus attention on the consequences, criteria, alternatives, feelings, and ratings students considered as they made their decisions.

All of the episodes can be used in conjunction with other learning objectives, concepts, or topics not included in the lesson plan. If you select alternative objectives, write other questions, in addition to the follow-up questions listed, before the class activity begins to ensure alignment of the follow-up discussion with your pre-activity objectives.

Follow-up Questions: Emphasis on the Decision-Making Strategy. These questions, suggested in Table 4 (page 20), help students focus and reflect upon the decision-making strategy and processes they used to respond to each episode. Put the list in a convenient place to use on a near daily basis—both in conjunction with these episodes and following any decision activities students complete. Ask one or more of the questions after students have made their personal and group decisions. These questions are particularly important for successful teaching of the steps of these decision strategies.

Merely because students follow directions and use a set of steps for a particular decision-making strategy does not mean they (a) know they are using the strategy, (b) know the steps in the strategy, (c) know the order of the steps, or (d) comprehend the steps or the strategy well enough to know when, where, and how to use it effectively in other situations. To teach for the acquisition and skillful use of decision-making abilities, you must help students internalize the effective use of these steps and strategies. Mastering the use of any decision strategy requires more than just having students use activities such as these. Each student must consciously and repeatedly consider the steps of each strategy and get information about how well he or she can use the strategy to make decisions. For these reasons, use an assortment of questions similar to those in Table 4 following every episode.

Students must be given sufficient, uninterrupted time to answer the questions selected. As often as possible, to enable them to better comprehend their own thinking processes and strategies, students should be allowed to write down or reveal their answers publicly. In this way, the students, their peers, and you can provide adequate and accurate feedback as each student tries to develop and maintain effective decision-making and problem-solving skills. Without such feedback, the acquisition and eventual transfer of these strategies and decision-making abilities to other life situations is not likely to take place.

Follow-up Questions: Emphasis on Behaviors as Members of Cooperative-Learning Groups. These questions, provided in Table 5 (page 36), help students focus and reflect on behaviors and attitudes used in responding as members of small groups to each episode. Ask one or more questions from this list after students have completed their work as members of groups. These questions are particularly important for successful teaching of behaviors and attitudes relative to effective and productive work as members of groups.

PURPOSES OF THESE STRUCTURED DECISION-MAKING EPISODES

Society and Science contains 33 episodes that may be used during whole-class instruction, in learning centers, as part of small-group instruction, as cooperative-learning activities, or as independent student tasks and assignments as appropriate to curriculum, student, and class needs. The episodes may also be used to complement or enrich a student's study of the subject matter in a course or program of study.

While examining the content, issues, and situations relevant to important science or technological events and data, students will be required (a) to gather, examine, and clarify empirical and relevant data; (b) to determine and consider alternatives; (c) to examine and consider the probable and possible consequences of each alternative; (d) to determine and decide the standards or reasons to be used for basing a decision; (e) to make decisions; (f) to decide on how their decisions may affect them or others; and (g) to evaluate the merits of their reasoning, decisions, and the consequences for themselves and others. Equally important, students will consider personal, cultural, social, technological, and scientific consequences that may result from their decisions.

The processing tasks in each episode have students rehearse and acquire processing strategies and logical reasoning abilities vital to making effective decisions. During these episodes, students will be working as individuals or as members of a group asking and answering questions, gathering relevant data, analyzing problem situations and dilemmas, interpreting and assembling information, considering rational and emotive reactions to problem situations, and drawing conclusions. Essentially, they must use the embedded decision strategies as they reason, contemplate, inquire, and make personally and socially meaningful decisions in response to the society-, science-, and technology-related situations and information.

Students can learn to be competent decision makers, just as they can learn to be competent at any other subject matter or process ability. Learning tasks that require the regular use of the same, specific decision-making steps and strategies will help students acquire and successfully use the strategies. When used appropriately, students can acquire sound strategies to help themselves cope with the problem-solving tasks they face in situations related to science and technology as well as in their everyday life. Furthermore, they are likely to continue to encounter situations relevant to the academic content and concepts as adolescents and adults.

WHERE TO USE THESE EPISODES IN THE CURRICULUM

The episodes include problems and situations related to interactions and relationships among society, environment, science, and technology. They vary in the amount of focus on each of these four areas as well as in the extent to which the interactions and relationships among these areas are made obvious. The focus of these episodes concerns issues, topics, and content usually emphasized in junior and senior high school classrooms.

There are many subjects and courses in which these lessons are appropriate. The episodes are easily adapted to different classroom settings and subject matter. They are appropriate for emphasizing student learning and comprehension of subject matter. They may be used to blend cognitive learning with affective and process-related responses, and used to help students develop problem-solving and decision-making skills.

The episodes may be used at different times during a unit of study. You may want to use a particular episode to introduce a new unit or new content area, as an integral part of a unit, or to help students summarize or consolidate a unit being completed. The episodes are nearly self-instructional, freeing students to become actively involved with the content and concepts while freeing you to facilitate those students and groups who need special attention.

These episodes can be effectively completed by small groups with a large-group (full-class) discussion following. This format helps break routine patterns of lecture-textbook-review-discussion. Using these episodes with small groups is an attractive option, since step-by-step instructions as to how students are to proceed are included in every short-story episode.

The episodes are easily adapted to different classroom settings, grade and age levels, and courses. They have motivated and informed students enrolled in regular and remedial classrooms. They have been used successfully in gifted education, advance placement, and teacher-training programs. All of these episodes, have been successfully field tested with students in middle and high school classrooms as well as among teachers during inservice workshops and methods courses.

Research Findings to Support the Use of Structured Episodes

These episodes were designed according to one model of constructing content-centered decision-making learning episodes (Casteel and Stahl 1992). To determine the effectiveness of this model, two studies of the effects of these episodes on students' subject-matter content retention and attitudes were conducted. Eighteen classes with 394 students participated in one six-week study (Stahl 1981). Experimental group teachers used one episode a week to replace their normal coverage of the subject matter. Control-group teachers taught their classes as usual. One and four weeks after the sixth episode was used, students in the 18 classes were given an unannounced 32-item content test and a 60-item attitude survey.

On both the immediate and delayed post-tests, students who had used the decision-strategy episodes had significantly higher content-achievement test scores and held significantly more positive attitudes than students who had not used such episodes. The content test included a majority of items aligned with subject matter information found both in the learning episodes and the textbook. The remainder of the items included subject matter information found only in the textbook. The attitude test contained ten major subscales (communications, problem solving, decision making, personal consistency, assenting-dissenting, acceptance of self, openness to content, openness to participation, and open-mindedness) to which the experimental-group students responded more positively than the control-group students. A second study involving over 400 junior high school students produced similar results for both subject matter achievement and positive affective scores (Hunt 1981). These research findings reveal that activities such as those in this book help students better comprehend and retain subject matter content while contributing to the production of significantly more positive attitudes. For more information on the details of both studies or the model used to develop these episodes, please refer to the references at the end of this book.

CHAPTER **4**

Using the Episodes in a Cooperative Learning Context

The episodes in this book can be completed as individual decision-making tasks or in a large-group setting. All may be completed within the context of group decision making.

By taking the steps outlined in this chapter, teachers can arrange the groups and group tasks to ensure cooperative learning occurs among students. In completing these episodes, students tend to cooperatively study must-learn content and skills. The episodes facilitate students considering, comprehending, and rehearsing important content *as individuals* and *in a group as a group,* and then reconsidering, reflecting upon, and re-using the content as they make individual and small-group decisions.

The tasks are structured to facilitate cooperation. Students become actively involved in learning the content during problem-solving activities where outcome-aligned worksheets are disguised within the context of a story and a problem to be resolved. To complete each episode, students make structured individual decisions before making a group consensus decision. The group consensus task is merely a device to get students to rehearse the important subject matter *content* contained in the episode within the *context* of a problem-solving situation. Consequently, it is not important *what* decisions students make as individuals or as a group because *the group decision task is a contrived situation to ensure that students attend in meaningful ways to the content, concepts, and abilities they are to learn.* What is important is that students engage in the group decision tasks, because these ensure rehearsal of the content they should comprehend, assign meaning to, and retain.

Embedded in the episodes are clear directions as to how students, as individuals and groups, are to proceed to complete the required tasks. In this way, once students are handed the episode materials, they can operate individually and as members of a decision team without constantly asking you questions.

This chapter provides an overview of components that move the episodes in the direction of cooperative learning. It then describes the essential elements of cooperative learning that you must plan for, implement, and monitor to move these episodes from *cooperating* group tasks to *cooperative-learning* group tasks. A scenario is given that describes how a typical episode might be used in the classroom. The last section addresses specific questions you might have in using these episodes as cooperative-learning activities.

COMPATIBILITY WITH COOPERATIVE LEARNING

The episodes contain seven major components that help make the structured tasks more compatible with what students need for working cooperatively. These components are critical elements of these structured, academically oriented short stories—stories that require cooperation prior to and during the decision-making process. This section describes the seven components.[1] (The first five components are not included in the exploration format episodes, since the exploration format episodes do not include content-based short story situations.)

1. *The episodes contain a large quantity of must-learn information that is directly aligned with the academic outcomes.* This component ensures that students encounter outcome-aligned content and concepts. Since the episodes often replace or supplement traditional ways to get students access to must-learn content and concepts, they must include some of the content students would be expected to acquire though textbooks, worksheets, and lectures. In addition, the plot and demands of the story have students consciously considering and reconsidering the content as they move toward a decision.

2. *The episodes contain a description of a problem-solving situation in story form that includes sufficient background information for comprehension of the context in which a number of options are to be considered and a decision made.* The characters in the story are required to resolve the problem by using one decision strategy to consider the options available. As the plot unfolds, each story makes explicit who the character (or group) is and at least one problem that must be resolved. The story provides a clear narrative within which students consider a situation, a set of options, and the must-learn academic content. It enables students to develop a perspective that would likely be taken by one or more of the characters; it provides them with a context in which to academically examine the context, content, and options; and it constructs a situation in which deciding among options is a meaningful activity and has implications beyond the decision task.

The story contexts enable students to move quickly into the roles of the persons in the story as they take students out of their present-day perspectives and move them toward the perspectives of characters in the story. These role-taking experiences increase students' consciousness and sympathetic perspectives of individuals in different roles and cultures.

Each story also includes one or more important constraints and limitations that characters must consider as they deal with the situation. These limitations may involve resources,

---

[1]The original model developed by Casteel and Stahl (1975) was refined (Casteel and Stahl 1989) and further modified (Stahl 1987, 1990, 1994) to make it more consistent with the requirements of cooperative learning. Both the refined and modified versions of the original Casteel-Stahl model are highly suitable for effective learning of academic content, positive affective consequences, and systematic rehearsal of transferable decision strategies. *Doorways to Decision Making* provides in-depth descriptions and how-to guidelines for constructing structured learning episodes. Workshops and sample activities are available from the author.

time, space, supply of information, or political or economic implications. For instance, a character may have only 1 hour to make a decision, lack the necessary equipment or personnel, or have too little money. These constraints help students understand that individuals in different times, places, and cultures are not free to make just any choice, because limitations reduce their freedom of choice.

3. *The episodes contain a list of options that the character(s) must consider.* Students make their decisions from among a set of options. In three of the five decision strategies, all the options are provided, and they cannot be changed, added to, or replaced. In these story episodes, the options are mutually exclusive—each represents a unique alternative such that two or more cannot be combined to form a new option. In other words, the characters are stuck with the options they are given. In these instances, the options are usually *homogeneous* in that they are (a) all possible policies, actions, criteria, laws, and consequences; and (b) about equal in their attractiveness, unattractiveness, and possible effectiveness. The episodes typically contain only options the characters really want or don't want, rather than provide some desirable and some undesirable options.

4. *The episodes contain clear instructions as an integral part of the story to inform students how and why they are to proceed to make their decisions.* Each episode has built-in guidelines as to how students are (a) to proceed through the activity, (b) to make their individual decisions, and (c) to move toward the group consensus decision. Once students get started in their groups, the story plot and details instruct them how to proceed to complete the episode. This structure allows the teacher to spend time monitoring the groups rather than constantly giving procedural directions.

Of equal importance, the built-in instructions allow group members to function interdependently as they talk with each other about where they are in completing the tasks, what they need to do next, and what they have to do as individuals and as a group to complete all the required tasks. Within their groups, students ask and answer procedural questions and guide one another on-task to complete the learning focus and work tasks.

5. *The episodes contain predecision task sheets, to be completed first by each student and then by the group.* Predecision task sheets reinforce the structure of the decision tasks. They tell students what to record as public evidence of their predecision considerations and choices. In most instances, the activities are designed to have students, through their active participation, consciously consider the critical information, conceptions, and abilities they are to learn. Typically each student completes a predecision task sheet, then the group works cooperatively toward making a consensus decision. These sheets provide clear, explicit directions for students. The more explicit the directions, the greater the number of students who will proceed on-task from sheet to sheet without asking questions about what to do next.

In some cases, students are asked to pause at key places in a story to review the story and to work with their group to check that all members comprehend the targeted data and context before moving on. In other episodes, students are asked to complete structured analysis tasks and discussion prior to completing one or more predecision task sheets.

6. *The episodes contain a set of final decision sheets to be completed, first by each student and then by the group.* Usually there is one form for students to record their final individual choices and a second sheet for the group's decision. These sheets provide explicit directions on what students need to report as public evidence of their preferences, reasons, and priorities. The personal and group deci-

sions themselves are rarely important for their own sake. The decision sheets provide closure for the story, practice for the required decision strategy, and allow for systematic reconsideration of much of the must-learn information.

7. *The episodes are followed by a set of questions as discussion starters to focus inquiry and learning.* These questions help students review what they have just studied and considered as part of the academic content and context of the episode. They enable students to assign meaning and find connections between what they did in their groups to other data; themselves; and present, past, or possible future times, places, situations, and people. And they enable students to compare the perspectives and decisions made in reference to the story characters with decisions they may make in the similar situations. If the questions are limited to the must-learn content, a fourth role for these questions is as test review questions.

8. *The episodes may also be followed by a set of questions that require students to consider the decision strategy used during the decision-making task.* These questions, which are universal to all episodes and listed in Table 4 (page 20), are to be used after students have completed a decision-making task. They help students articulate, reflect upon, and consider the content, steps, context, and results of using the decision strategy built into the episode.

ENSURING COOPERATIVE LEARNING REQUIREMENTS ARE MET

Learning episodes designed with the seven components described above go a long way toward ensuring that the required elements for cooperative learning are met. *The other essential elements for cooperative learning are met only as you ensure the following are completed:*

A. Positive Interdependence

You can structure for positive interdependence by (a) making it clear that all members of the group are to master the targeted content and abilities; (b) informing students that part of their personal effort is to help every member of their group learn the content and abilities targeted; (c) providing group rewards (for example, if all members of a group score or average 85 or higher on a test, all members earn 10 bonus points for prizes or rewards outside the unit or term grades[2]); (d) assigning specific complementary roles to each group member (for example, one student is the Checker, another the Elaborator, and another the Recorder); and (e) dividing the resources equally among all members. You are encouraged to find other means to ensure students work to help every group member succeed. The positive interdependence requirement is met only when all members of each group feel they *sink or swim together* (Johnson, Johnson, and Holubec 1993).

To meet this requirement students will need sufficient, uninterrupted time (a) to stop at appropriate moments in an episode to review systematically the context, content, and situation; (b) to work cooperatively to attain a group decision (for example, as a group of scientists in a class); and (c) to review each others' decisions for the predecision task to provide group reactions and clarifications followed by their mutual efforts to reach a group decision. For instance, in most episodes, students cannot make a group decision until each has made an

[2]Bonus points within the cooperative learning approach does not necessarily mean additional points to be added to each student's grade; they can be earned toward rewards or prizes outside of unit or term grades.

individual decision that in turn cannot be made until all students in the group have cooperatively and systematically considered the content, context, and options in the story.

B. Face-to-Face Interaction

Students arrange themselves so that, from the beginning to the end of their work though an episode, they are positioned and postured for close, face-to-face discussion. They are encouraged to use "12-inch voices" so they can interact comfortably while keeping the talking in the room at an acceptable volume.

C. Individual Accountability

Students must eventually be formally tested on the extent to which each has learned the targeted academic content and skills embedded in the episode. For example, in the episode "Bo or Zo?" students would be expected to provide a number of the specific names of the several branches or subdivisions of botany and zoology, describe the major activities of each branch, and describe the equivalencies and differences between botany and zoology as fields of scientific endeavor. This element is enhanced when you clearly state, while introducing the activity, exactly what information, concepts, and skills students are to learn. You may want to compute average improvement scores per team and recognize teams for their collective achievement.

D. Heterogeneous Groups

To meet this requirement, organize three-, four-, or five-member groups so students are mixed as heterogeneously as possible according to academic abilities, ethnicity, socioeconomic levels, and gender.

E. Equal Opportunity for Success

Cooperative learning intends to strengthen the academic learning of each group member. You put students in cooperative-learning groups so that, eventually, all students will be more successful, academically and as individuals, than they would be by studying alone or in non-cooperative-learning groups. Therefore membership in any group should provide each student with an equal chance of learning what is to be learned.

F. Positive Social Interaction Abilities, Behaviors, and Attitudes

To complete the group tasks, students must work together *in a group as a group*. However, just because students are placed in groups and expected to use appropriate social and group abilities does not mean they will automatically use and improve these abilities. As students work through structured cooperative-learning activities, they must engage in leadership, compromise, negotiation, and clarifying. With sufficient rehearsal in using each ability correctly, students may become, over time, skilled users of these abilities. Students may enter the classroom as "novice" users of these abilities and leave as "skilled" users. You may need to describe in specific language the social interaction behaviors and attitudes students are expected to use and master, including those for leadership, trust building, communication, conflict management, constructive criticism, and encouragement. You may assign different students particular group roles to ensure they consciously work on these behaviors in their groups.

As students work through these activities, they also tend to become tolerant of diverse viewpoints, to consider others' thoughts and feelings in depth, and to seek more support and clarification of others' positions. It is not uncommon to have students of different academic or ethnic backgrounds or gender deal with critical issues or dilemmas in ways rarely found in other instructional strategies.

G. Clear, Specific Student Learning Outcome Objectives

You must start your planning knowing specifically what information and abilities your students must learn and be able to use well beyond the end of the group task and curriculum unit. These outcomes may include academic content abilities, cognitive-processing abilities, task-completion abilities, and social-interaction abilities. Cooperative learning and cooperative-learning groups *are means to an end rather than ends in themselves.* Selecting clear, specific outcome objectives before the groups form and begin their cooperative tasks keeps groups focused on what is to be learned rather than merely on what students are to do to complete the activities.

H. Mutual Ownership of Student Outcome Objectives

Students must accept the targeted outcomes describing what they are to learn as being their own. They must accept that everyone in their group is expected to learn a common set of information and abilities.

I. Sufficient Time for Learning

Each student and group needs enough time to learn the targeted information and gain the abilities to the extent expected. If students do not have enough time, the academic benefits of cooperative learning are limited (Stahl 1992).

J. Public Recognition and Rewards for Group Academic Success

Arrange for individuals and groups who meet or surpass high levels of achievement to receive rewards in formal public settings. Compute average scores per team and recognize teams for their collective achievement.[3] You can establish a class awards presentation for outstanding performance or set up a permanent bulletin board display along with a public announcement. You can also invite outside professionals, such as the principal or department chair, to attend an awards ceremony. The awards will vary with the critical criteria and must be something valued by students. Awards should be given as soon as possible after the individual test has been given.

K. Systematic Reflection on Each Group's Processing Efforts

Students need to spend time after each episode to reflect on how they worked as a team to consider the critical content; to help each other comprehend the context, content, and decision strategy; to progress toward a consensus decision; and to maintain positive behaviors and attitudes for everyone's success. To ensure this reflection occurs by design rather than by chance, you should provide a structured reflection task and sufficient time after groups share their responses.

The questions in Table 5 (page 36) can guide students in examining their work, behaviors, and attitudes concerning their membership in groups. An assortment of questions similar to those in Table 5 should follow each episode where small groups are used. These are not built into the episodes and depend on your implementation. Use the lines to the left to check the specific questions you will ask for a particular episode. Use the blank lines to include other questions you want to ask.

---

[3]Excellent examples of how to analyze group grades and award improvement points based on individual and team achievement are in Chapter 5 in Stahl (1995a, b).

Table 5
Behavior and Attitude Focusing Questions

Use these questions to focus and guide reflection of behaviors and attitudes that students used, were to use, and did not use while working as members of a learning team.

_____ **1.** What behaviors were most effective in helping your group successfully complete its task?

_____ **2.** What behaviors worked against your group's success in completing its task?

_____ **3.** What words were most effective in keeping your group on task?

_____ **4.** What words were used to get your group off task?

_____ **5.** To make sure your group stayed on task, what things could members of your group have said?

_____ **6.** What behaviors were effective in helping you learn what you were suppose to have learned?

_____ **7.** What worked best to help members of your group learn?

_____ **8.** In your group, how hard was it to persuade others to accept your point of view?

_____ **9.** To make sure the group worked as a team, what did you have to do?

_____ **10.** Suppose you wanted to help your group succeed in future team tasks. What are three behaviors that you would most want to see team members doing in the group?

_____ **11.** Suppose you wanted to help your group succeed in future tasks. What are three behaviors you would *not* want to see team members doing in the group?

_____ **12.**

_____ **13.**

_____ **14.**

_____ **15.**

_____ **16.**

The use of *these elements almost always distinguishes non-cooperative-learning group work from cooperative-learning groups.* You will need to take steps to ensure these requirements are met for each group activity you use. More importantly, unless these elements are used frequently and correctly for prolonged periods of time, you should not expect to achieve the many positive results that can be achieved for cooperative-learning tasks.

A CLASSROOM SCENARIO

What does this strategy look like in the classroom? This section uses a classroom scenario to illustrate what would occur in a "typical" classroom situation where this strategy and a relevant episode are used.

Ms. Marilyn Fischer wants her students to comprehend a number of ideas associated with the major accomplishments of the space program and to consider the implications of these accomplishments on life on this planet. She strongly feels her students need worthwhile data concerning space activities and explorers. She wants them to reflect upon the meaning of these accomplishments and of their possible consequences on other space missions and the society in general.

To help her accomplish her goals, Ms. Fischer locates the structured resource "Off We Go" (page 189) that contains fifteen of these accomplishments in the form of a short-story episode. She notices that the space achievements are listed in a context that requires students to consider each mission and achievement on its own merits. Supplementary information about the space program, its characteristics, and problems and concerns about space travel will be provided later in the unit.

Following the introduction of the unit, she overviews the background of the space program, emphasizing early space missions that have often been linked to the "race for space" between the United States and Soviet Union from 1957 to 1989. She points out some features of the American space program with special emphasis on the flights carrying humans into space. She tells students they are to learn details surrounding the achievements of humans in space, including but not limited to missions with humans aboard, and to be able to state reasons for such space missions and achievements. Finally, she tells students they are to learn consequences of these missions on life on this planet, including consequences on science, scientific research, and technology.

After a few questions to clarify certain points of this background information, she assigns students to heterogeneous groups of four. She asks students to think of themselves as members of the Brownstone University selection committee, and she directs them to spend a few minutes reviewing their notes, especially those referring to the background of the space program between 1945 and 1957. Ms. Fischer allows students 10 minutes to assemble their groups and review the background information. She then hands every student a copy of "Off We Go."

Ms. Fischer directs students to read the story and to fill in by themselves the requested information on each sheet, as completely as possible. Then the entire group will work together to share responses, clarify ideas, and ensure that all group members have understood the information and completed the forms. She tells them to make individual decisions before going to the group and not to complete a new form or worksheet until the entire group has worked together on the current sheet. Finally, she directs them to reach their final group decision by consensus. They are not, she emphasizes, to make their group decision by majority vote. In response to two questions, she describes what it means to reach a vote by consensus. For this activity, she decides not to select group leaders but to allow the groups to take on their own leadership patterns.

To emphasize the importance of their participation and study, she states that many of the important details of the space program, its achievements, and its consequences they are to learn are included in the episode.

The Brownstone University selection committee members then begin reading the story and completing the forms in the handout.

Students read the story, some pausing along the way to ask a group member the meaning of a word or for more information about other space-relevant activities during this era. As they get to the predecision task sheet, each student writes the information as directed, frequently turning back to the story for details. Students in each group wait until everyone is done or about done.

Then each group reviews what was written by each individual and works on clarifying, correcting, eliminating, and adding to what each member wrote. More than one student in each group again refers to the teacher-given notes or to the story to confirm a response or get information to support a decision. Ms. Fischer, moving about the class, is not surprised that nearly every student is on task. She is pleased with students' comments about reasons why certain space missions were conducted.

Students proceed with the episode, eventually making a personal decision in the context of the story as to the five missions they most want to recognize with museum displays and the five achievements they are most willing to give up. They individually determine the reasons for their choices. Time and again she hears references regarding students' individual and group tasks. They remind one another of the particular directions in the episode that inform them what to do and when to do it.

Now that their personal decisions have been made and recorded, the groups spend time trying to reach a consensus on the decisions. Individuals take time to give reasons for one choice over another, frequently concerning possible consequences they consider to be important or not so important. Ms. Fischer has to remind one group that they cannot place all the achievements in the museum: "There is not enough space in the museum to adequately portray every one of the fifteen missions," she whispers to a student in that group. In another group she hears three students inform a fourth that they cannot include a mission that is not listed.

After 20 minutes of group effort to reach a consensus, Ms. Fischer tells the groups to stop their discussion. (Ms. Fischer is aware that final decisions are not important here since the consensus decision task is only an opportunity for students to reconsider the important details as they move toward their decisions in the public forum of their group discussion.)

Each group takes a minute or two to announce to the class what decisions it made and the reasons behind them. They also mention how their group operated in making their decision. Facts and notions about the space missions and their consequences are clarified or corrected.

As the period comes to a close, Ms. Fischer provides a focused lesson closure, emphasizing the important areas of information students are to remember. She congratulates students for working so well in their groups.

At the start of the next class period, she conducts a short review and then presents new information to supplement that in the episode. Students respond to ten questions Ms. Fischer has selected to guide their reflection of the situation they have studied. She decides not to review the decision-making strategy students used during the episode or the positive and negative behaviors they exhibited during their group work. Ms. Fischer expects to do this after students complete each of the next four episodes planned for the unit.

Over the next three days, Ms. Fischer asks four or five questions each day concerning the space program, the fifteen missions, their achievements, and their consequences.

At the end of the unit students take individual academic tests aligned with the outcome objectives targeted for the unit.[4] They are tested on their retention of details about nine of the space missions described in the episode, a number of reasons why these space missions were made, and consequences of these space achievements for individuals, societies, science research, and technology on earth.

This scenario is a frequent scene in classrooms where these episodes are used. Once students are organized into groups and begin reading the episode, you move into the role of monitoring and assisting student progress through the steps described in the short story. Of course what occurs before and following each episode may vary in length of time or depth of information, thus allowing for individual teacher differences, the needs of students, and the content being studied.

SELECTED QUESTIONS

What Areas of Student Learning Outcomes Are Appropriate for These Episodes?

From personal observations and extensive feedback from teachers who have used episodes based on the Casteel-Stahl model, appropriately structured episodes tend to lead to at least the following positive results:

- Increased comprehension of academic content, concepts, and processing abilities
- Increased retention of content-specific information
- Increased voluntary participation in academic tasks by slower learners as well as students from all backgrounds
- Improved attitudes about the value of the academic content, concepts, and skills being studied in the course
- Improved abilities to work with others within a group setting
- Increased time spent on task to complete the academic episodes, in contrast to the time students typically spend reading textbook and other reference material that stress the same content
- Increased interest in the subject matter and in participating in the class
- Improved self-image of students as students
- Increased on-task talk among students in small groups
- Increased on-task rehearsal of a number of social skills and group interaction skills necessary to function in any group as a cooperating, effective member
- Increased opportunities for leadership experiences

[4]Chapter 5 in *Cooperative Learning in Language Arts: A Handbook for Teachers* (1995) and *Cooperative Learning in Science: A Handbook for Teachers* (1995), both edited by R. J. Stahl, provide details concerning how a teacher may assign improvement points for individual and group achievement as a result of student involvement in these episodes. Details may be obtained by writing the author.

At What Levels of Schooling May These Episodes Be Used Within the Context of Cooperative Learning?

These small-group decision episodes may be used from upper elementary through graduate-level university classes and in-service staff development environments. Of course, such factors as the complexity of the topic chosen, the abilities of students (for example, their reading and writing abilities), and time available will affect how well students will complete an episode in any particular course at a given grade level.

How Much Class Time Is Needed to Complete an Episode Within the Context of a Cooperative-Learning Strategy?

Major factors affecting the time required for successful completion of an episode include (a) breadth or difficulty of the content to be learned; (b) the length of the story to be read; (c) the number of predecision and decision task sheets to be completed and discussed; (d) the amount of time students need to complete their discussions; and (e) the number of students per group. An episode could extend from as brief as one class period to three or four days. Even though you may be very familiar with this instructional approach and students are accomplished in the use of such episodes, the time students may need on a particular episode can still be less or more than what you might expect.

* * *

Not all decision-making activities are created equal. Few will generate the positive outcomes likely through these episodes. Many of the episodes, especially when used within a cooperative-learning context, are unique in that nearly all of the elements of the strategy are manifested in the episodes. They are designed so students, as individuals and as members of a group, complete numerous cooperative tasks as they work through the short stories. When accompanied by heterogeneous groups, sufficient time, supportive teacher monitoring, and appropriate follow-up activities, these episodes generate results compatible with many widely used cooperative-learning strategies.

The structure and components for the episodes in this instructional approach evolved over years. More years were spent refining these components so nearly all students who complete the episodes must share, cooperate to help one another comprehend and learn, work together as a group, and learn the targeted academic content. In effect the episodes are designed to guide cooperative-learning tasks as students progress to attain a collaborative consensus solution to a contrived problem.[5]

---

[5]Nearly all episodes following this strategy may be completed without students working through all the predecision task sheets or by having students make group decisions by majority rule than by consensus. While this is true, in doing so the teacher moves away from the strong tie-in to cooperative learning and converts these episodes into activities where students primarily cooperate to complete the assignment. While many positive results do occur in such situations, the most powerful and enduring results occur as these episodes are used as provided, along with teachers ensuring the many other elements of cooperative learning are included.

Forced-Choice Decision Episodes

Forced-choice decision episodes give students a problem, dilemma, or situation in which a major character or group must make choices from a limited number of almost equally desirable, nonchangeable alternatives. Only one option can be selected and, once selected, the remaining options are no longer available. Learning tasks emphasizing the use of this strategy allow students to practice making decisions under forced-choice circumstances and help them understand the factors that must be considered in such restrictive situations.

THE DECISION EPISODES

MY FAVORITE MONSTER

One month ago, Betsy Allen raced through the door waving a telegram from the National Science Society. Her family eagerly listened as Betsy read, "Based on your letter, you are one of five students selected to accompany scientists on an expedition to Africa. You will investigate rumors of a creature twice the size of an elephant. These rumors suggest this creature is a dinosaur. Details to follow in the mail!"

Now, a month later, she is leaving for a remote area of West Africa. The area includes miles of uncharted jungle. The natives, primarily pygmies, have reported having seen a huge, water-dwelling monster roaming the jungle and rivers. Visitors to Africa have heard many rumors about this creature. In the last several months, reported sightings by the natives have increased significantly.

Betsy received information in the mail about the other students and scientists. Aric Beals, a zoologist, is the scientist who sent the most information. His detailed notes included a color drawing of the hippopotamuslike creature seen by the natives. He reported that the brownish-gray creature is usually seen in the early morning and late evening. At these times, it comes out of the water to eat plants along the river. At other times, the creature seems to stay in the water. The footprints found by the river are about 36 inches across and similar to an elephant's footprints except that they show claw marks. The creature's estimated weight is 6 to 15 tons. Its short, thick legs support a body stretching about 35 feet from its small head down its long neck to the end of its tail. Aric reported that the natives had identified pictures of elephants and hippopotami. All who have seen the creature swear it is not an elephant or a hippo.

When the witnesses were shown pictures of the vegetarian (plant-eating) dinosaur apatosaurus (commonly called brontosaurus), they identified it as the creature they saw. The scientists agree that it is unlikely an apatosaurus still roams the jungles of Africa. But, if this creature is in the dinosaur family, members of the expedition need brief descriptions of likely dinosaurs they might encounter. The team of experts and scientists from the National Science Society prepared information on six types of herbivorous dinosaurs. Their report is attached.

Betsy is fascinated by the possibility of discovering some form of prehistoric life. Scientists know something about prehistoric animal and plant life because of fossils they have found. They have tried to imagine what these animals did and looked like by the fossil remains. They have no actual skin samples to examine, but they think the skin had brownish-green tone similar to today's lizards.

Imagine you are Betsy Allen and that you will take the trip to search for the creature. Before going any further, review the story so far. Be sure you can answer the following questions.

1. What are two major purposes of this expedition to Africa?

a.

b.

2. From the details of this story, what evidence is there that a large creature may exist in the jungles of Africa?

3. What are five important facts that describe the place in Africa where the expedition will occur?

a.

b.

c.

d.

e.

4. What does this story have to do with science?

With scientists?

With what is important to people like you?

5. For what three reasons should a modern-day society or nation spend money, time, and effort to look for a prehistoric creature on another continent?

a.

b.

c.

Share your answers with members of your group. Revise, correct, and add to your answers as needed. Make sure all members of your group have adequate answers to all of these questions before going any further.

Now read and study the attached summary report. Write answers to the questions on Predecision Task Sheets 1 and 2. You will then be asked to make a number of important decisions.

Summary Report on Herbivorous Dinesaurs

Prepared by the National Science Society for the Expedition to Africa in Search of the Unknown Dinosaurlike Creature

It is believed the last dinosaur died several millions of years ago. It is unlikely, according to scientists, that any dinosaur still lives in today's world.

The word *dinosaur* comes from two Greek words meaning "terrible" and "lizard." This name was given because many of the first fossil skeletons found of dinosaurs indicated that they were large, powerful, and fierce carnivores (meat eaters). Dinosaurs were not lizards, but they were reptiles just as lizards are reptiles. Also, a great many dinosaurs seemed to have been gentle creatures. Scientists believe dinosaurs roamed the earth for a hundred million years, then vanished suddenly and quickly. There is no widely accepted evidence as to why they disappeared.

It is possible the large creature rumored to live in Africa is one of the six herbivorous (plant eating) dinosaurs described below. The society asks that you study these descriptions carefully.

1. Brachiosaurus

Until the fossil discovery in Utah of the supersaurus, the brachiosaurus was believed to be the largest herbivorous dinosaur. This long-necked animal was 80 feet long with a height of 20 feet at the shoulders and up to 40 feet at its raised head. It lacked the muscles to rear up on its hind legs, but had high shoulders and an elongated neck and could reach very high leaves and twigs, much like the modern-day giraffe. This slow-moving, harmless creature is estimated to have weighed about 80 tons, and its brain is believed to have been the size of a kitten. Because the enormous nostrils were located in a bony ridge at the top of its skull, the brachiosaurus could be nearly submerged in water and still breathe.

2. Apatosaurus

The apatosaurus is the most widely known of the plant eaters and is believed to be among the gentlest of all prehistoric animals. It is nicknamed the "thunder lizard" because when scientists first saw the enormous size of its bones, they thought it must have sounded like thunder every time its foot hit the ground. The adult apatosaurus weighed 35 to 50 tons and was 70 feet long. Its small head allowed it to swing about and eat low vegetation or to stretch its neck to feed on the tree tops. It is believed this animal was constantly eating. It is believed that its trumpeting sound could often be heard as it fed in shallow waters or on dry land near a swamp. The tail of the apatosaurus was long and narrow and could be used as a whip. This dinosaur's skeleton was among the first found by scientists.

3. Yaleosaurus

The yaleosaurus probably lived before the brachiosaurus. It is believed to have walked on its hind legs as well as all four legs. There were stripes down its back. This creature was only about 8 feet long and somewhat resembled the tyrannosaurus, the most famous meat-eating dinosaur. The small size of the yaleosaurus may have allowed it to run from the meat-eating dinosaurs. Discovery of this dinosaur may give information about the first, small dinosaurs that later became the huge creatures of 70 to 100 feet in length.

4.	**Plateosaurus**	The long-necked plateosaurus was about 20 to 26 feet long and weighed 1,500 pounds. It stood 12 feet high when rearing up to eat leaves and tree tops. It, too, might have walked on two or four legs. This plant eater had short forelegs with four fingers and a big, clawed thumb with which to pull down vegetation or fight enemies. It is imagined this creature had green splotches and stripes on its back to help hide the creature from its enemies. If this creature is discovered, it could provide facts on the development of herbivorous dinosaurs because it is between the 8-foot-long yaleosaurus and 70-foot-long apatosaurus.
5.	**Camarasaurus**	The camarasaurus was nicknamed the pygmy of the amphibians. The largest fossils found were only about 40 feet long, half as long as the brachiosaurus fossils. It is the smallest of the large, four-footed herbivorous dinosaurs. The size of its skull is similar to that of a modern-day bulldog.
6.	**Diplodocus**	Because its tail and neck were very long, thin, and about the same length, some scientists imagined this "double beam" dinosaur was much like a seesaw. Its tail and neck seemed to balance each other. Its length of 90 feet is one-third the size of a football field. However, it weighed only 12 to 20 tons. Imagine, it was more than 80 feet from the diplodocus' brain to the tip of its tail.

It is unlikely the dinosaur is a supersaurus. Discovered by Dr. Jim Jensen at Dry Mesa quarry in Utah, the supersaurus is the largest dinosaur known. The few fossil bones suggest that it was at least 100 feet long, towered 50 feet high, and weighed about 100 tons. Its weight is estimated to have been 50 times more than that of a modern-day elephant. If estimates are correct, the supersaurus ate as much as 500 to 800 pounds of vegetation each day, spending 18 hours per day eating. Such a monster could not go unnoticed in today's world.

All of these dinosaurs may have used their tails as weapons. Fossilized footprints show pads on their feet similar to an elephant. Some of the footprints show toes with claws that probably helped keep the creatures from sliding in the mud along the riverbanks. It is assumed these animals spent a great deal of time in the water to help support their enormous bodies. Their teeth enabled them to chew only soft plant foods, especially leaves. Scientists today are trying to find out whether these dinosaurs had small true brains or no real brain. Some scientists speculate that the longest dinosaurs may have had two brains, one in the skull and one in the hip area or the end of the tail. The brains of these creatures may have been very different from the brains of today's large reptiles and mammals. The spinal cord may have had special nerve cells, called *ganglia,* that controlled the legs and tail.

While we have a great deal of information on these dinosaurs, most of what is written about them is the result of educated guesses from what scientists believe to be true. Finding a live dinosaur would do much to help verify our thinking about them. Such a discovery could be used as physical evidence to support or discount many of our "guesses" and beliefs about dinosaurs.

PREDECISION TASK SHEET 1

Now that you have read and studied these dinosaurs, take time to review what is "known" about each. Then write at least two important facts about each one in the space below. In addition, write the best reason for finding each prehistoric dinosaur as a unique animal. Your reasons must be different for each dinosaur.

Name and Description	Best Reasons for Finding This Dinosaur
1. Brachiosaurus	**a.**
	b.
2. Apatosaurus	**a.**
	b.
3. Yaleosaurus	**a.**
	b.
4. Plateosaurus	**a.**
	b.
5. Camarasaurus	**a.**
	b.
6. Diplodocus	**a.**
	b.
7. Supersaurus	**a.**
	b.

After you have completed your individual answers for the complete chart, share your responses with the others in your group. Add to, revise, or correct your responses as needed. Add at least one reason for each dinosaur. Make sure everyone in your group comprehends the information before going on to the next page.

Imagine you are now in the jungle of Africa near the spot where the creature was last reported seen. There have been no physical signs of the creature, but strange sounds have been heard. This afternoon a member of the team said only a monster could make such noises. What kind of creature would be found?

As you sit around the campfire, you and other members of the scientific team begin to discuss the creature. You share information about all six dinosaurs you have studied. Each person begins to talk about which of the six dinosaurs is his or her favorite. There is excitement as each begins to guess which dinosaur would be the most fun to discover.

At this point, Reid Nance, the zoologist who knows so much about dinosaurs, stands up. All eyes point to him as his shape darkens the light of the campfire.

"Let's pretend," Reid starts, "that in fifteen minutes, one and only one of these six dinosaurs will come walking into this camp. Imagine that this will be the same kind of dinosaur we may find on our expedition. Of these six dinosaurs, which one would you want to see walk into our camp? Which one would the team as a whole want to discover on this expedition? Before you decide which animal, I want you to think about these creatures and the role of scientists."

Use Predecision Task Sheet 2 to consider the information available prior to making your choice. You may want to share your responses on Predecision Task Sheet 2 with others in your group before you make your final decision. Use the Individual Decision Sheet to structure and record your personal final decision. Use the Group Decision Sheet to structure and record your group's final consensus decision.

PREDECISION TASK SHEET 2

The questions below concern a number of ideas you should consider before making a decision as to what dinosaur you would like to observe on this expedition. Write answers to these questions individually. Only when everyone in your group has written personal answers should you share your answers with all members of your group.

1. What are two good reasons why a scientist would be interested in finding a live dinosaur?

   a.

   b.

2. To this point in the story, the word *monster* has sometimes been used to describe this creature and dinosaurs. In your own words, what is a monster?

3. Given your definition of *monster,* what are two reasons why the dinosaurs as described in this story would or would not be monsters?

   a.

   b.

4. Suppose your favorite dinosaur did walk into your camp. What information would it provide us about prehistoric life?

5. What are three important facts a live dinosaur would provide that fossils cannot?
   a.

   b.

   c.

6. If you were the first to actually see a dinosaur, what words would best describe your feelings at that moment?

7. Would you be a scientist just because you saw the creature? What are your two best reasons for your answer?
   a.

   b.

8. Suppose the expedition did not find a dinosaur. Would you still be a scientist? What are your two best reasons for your answer?
   a.

   b.

9. Suppose the dinosaur turns out to be carnivorous. What different information might this type of animal provide you?

Remember, after all members of your group have completed personal answers to these questions, share your answers with them. Add to, revise, and correct your responses as needed. Try to add correct information to that you have already written. Do not go any further until all members of your group comprehend the information as answers to these questions.

INDIVIDUAL DECISION SHEET

You and your group are to imagine you are the scientists and students on the expedition. You know some facts about six kinds of dinosaurs. You believe the creature you may find in Africa is one of these six kinds.

At this time *you* are to select the one dinosaur *you* personally want most to be discovered. This will be your "favorite monster." If there are other dinosaurs in the area, they are likely to be the same kind as the one that is found. Of the six dinosaurs listed below, put a circle around the kind you would want to discover.

Brachiosaurus	**Camarasaurus**	**Diplodocus**
Apatosaurus	**Plateosaurus**	**Yaleosaurus**

1. I selected the ____________________ as "my favorite dinosaur" because

2. This dinosaur would help me discover or confirm these four things about dinosaur life:
   a.

   b.

   c.

   d.

3. My least favorite creature is the ____________________ because

DECISION SHEET

What creature does your entire group want to see discovered? Discuss the six dinosaurs with your group and come up with the one dinosaur all of you would most like to discover. You will need to talk about each dinosaur. Make your group choice without voting. Reach a common choice all members of your group will accept and support. Put a line under the name of the dinosaur to mark your group's choice. Then arrive at a group answer for each of the questions that follow.

Brachiosaurus	**Camarasaurus**	**Diplodocus**
Apatosaurus	**Plateosaurus**	**Yaleosaurus**

1. We selected the ____________________ as our "favorite dinosaur" because

2. The four most important facts this dinosaur would help us confirm about prehistoric dinosaur life are
   a.
   b.
   c.
   d.

3. Our least favorite creature is the _________________ because

4. As a member of this group, I can accept our group choice because

REVIEW AND REFLECTION QUESTIONS

Suggested follow-up questions to focus and guide inquiry and learning.

a. According to the story, what are the names of six types of herbivorous dinosaurs?

b. From the story, what are four characteristics of most herbivorous dinosaurs?

c. What does the term *prehistoric* mean?

d. Why would a geologist be interested in new information on prehistoric life?

e. What did the creature that had been seen in a remote area of West Africa look like?

f. What are three plausible explanations for why dinosaurs disappeared?

g. If a dinosaur is sighted and captured in Africa by scientists from a non-African nation, who would own the dinosaur?

h. Suppose a dinosaur does exist in the jungles of Africa. Should it be captured?

i. Where are at least three locations where dinosaur fossils have been found?

j. If you were one of the natives, how would you have felt to discover there really was a dinosaur? If there were not a dinosaur?

k. Within the last year, what "expeditions" have you been on?

l. Why are people in modern society fascinated by dinosaurs?

m. Some scientists consider the modern-day cockroach to be a *prehistoric animal.* To support their conclusions about the cockroach, what evidence would scientists use?

n. In what ways is the study of dinosaurs history? Science?

o. In what ways might new information about dinosaurs be useful to people today?

p. Suppose someone said gaining more information about dinosaurs is a waste of time, money, and effort. How would you respond to this claim?

q. Suppose a friend of yours was not interested in studying dinosaurs. What could you do to help him or her become interested in them?

r. What are the names of at least three modern-day descendants of the original dinosaurs?

s. What is it about dinosaurs that most fascinates you?

t. Photographs of the so-called Loch Ness monster appear to show features characteristic of certain types of dinosaurs. How could a dinosaur still be alive today?

u. A number of scientists recently claimed dinosaurs were warm-blooded rather than cold-blooded. If this were true, what other ideas about dinosaurs might we also have to change?

v. If this theory of warm-blooded dinosaurs is supported, what difference will it make in our present-day world?

w. If this theory is supported, what attitude will you have toward the scientists who claimed for over a century dinosaurs were cold-blooded?

x. If this theory is supported, will the scientists who supported the cold-blooded theory feel excited, disappointed, or upset?

REMAINS

You are a member of a special committee. The committee's task is to decide what will be done with the bones and artifacts of Native Americans that are in museums and the university in your community.

The three city museums have a small collection of Indian artifacts and two well-preserved full Indian skeletons. These skeletons, one of an infant and the other of a boy estimated to be between the ages of 9 and 11 years, are displayed in a glass case. The university has the partial remains of several hundred Indians and 13 full skeletons. It also has an extensive collection of Native American artifacts on display and in storage for scientific research. Oddly enough, the collections in both the city's museums and university were begun in 1876, the year of Custer's Last Stand.

Many of the bones and artifacts in one city museum are dated as being over 1,000 years old. Not one member of any present-day

Native American tribes is believed to be a direct descendent of the peoples whose artifacts and bones are housed in this museum. However, the largest collection can be traced to distant ancestors of present-day tribes.

The city's museums are not well known, but visitors have appreciated the Native American collection. The displays are popular among children. In addition, each year several thousand students walk or are bussed to the museum as part of their study of Native Americans. For most of these students, the displays are their only source of primary information on these peoples and their ways of life.

The collection at the university is well known. The largest part of the collection contains remains and artifacts of a tribe that vanished at about the time Caucasians entered the territory. One theory is that the entire tribe was killed by the early explorers or died as a result of diseases the explorers brought with them. Other theories include severe drought and an invasion by other warlike tribes.

There is no known direct connection between this now-extinct tribe and tribes that continue to live today. Because of the uniqueness of this collection, anthropologists from all over the world come to study the bones and artifacts. In addition, the university has the world's largest public exhibit for this vanished tribe.

With few exceptions, the remains in the city's museums and the university were found on land owned by private individuals. Almost nothing was found on government or public land. Furthermore, no remains were found using government money to finance the excavations. The museum and university collections are supported by funds from private companies and individuals.

These collections have come under attack in the past few years. As a member of the special committee, you are aware of the issues that have been brewing over these remains.

Native Americans have long contended that Indian remains should be left in the ground where they were buried. They say that unearthed bones and artifacts must be taken from museums and universities and returned to the ground. The tribes should receive these bones and artifacts for reburial.

Meanwhile, scientists, many of whom are archaeologists, claim that these objects are needed for study.

In 1990, the *Native American Grave Protection and Repatriation Act* was passed by Congress. This law requires museums and federal agencies to take inventory of their Native American collections and determine which Indian tribe or group each object came from or belonged to. Then, upon request, the bones are to be returned to the appropriate tribe. This is also true for religious objects.

The law sets up rules and guidelines for excavations on federal and tribal lands. A permit will be needed for digging. Artifacts may be excavated or removed only after the lineal descendants or the proper Indian tribe have given their permission. Anyone who discovers Indian remains, sacred objects, or artifacts must notify authorities immediately and make all reasonable efforts to protect the find. Stiff penalties are set for all violations.

This law seeks to ensure that Indian burial sites are shown the same protection and respect as non-Indian sites.

For decades, Native Americans have argued that they have been treated as second-class citizens by white people. Digging up and removing bones and artifacts show gross disrespect of the Indian and their Indian way of life. The acts violate Indian traditions and religious beliefs. Many Indians view the keeping of collections of bones and artifacts in museums and universities as a reflection of the belief that the white race is superior to the Indian race. To other Indians, these collections represent an extension of an old white expression, "the only good Indian is a dead Indian."

Anthropologists claim these bones and artifacts are needed to study the life and culture of Indians. Since these Indians had no written language, we can only know about them by studying these remains. Anthropologists argue that scientists and the public would know far less about ancient Egyptians without the study of their mummies.

Laboratory records, such as those compiled before returning bones and artifacts to a particular tribe, are not as useful as having the actual object to study. You cannot do chemical analysis of a bone fragment by analyzing a photograph of the fragment. You must have the actual bone. In other cases, new techniques for comparing individuals in the same tribe, individuals across different tribes, and individuals in relation to different time periods often require a reanalysis of the bone or artifact. Such comparisons and analyses would be inadequate and often impossible without access to the bones and artifacts. Scientists claim we would learn little new information about past Indian life and cultures were these remains reburied and no new bones or artifacts were allowed to be examined. Without written records, skeletons and artifacts such as these are an invaluable and irreplaceable source of data about the life of the first Americans.

Dennis Hastings, a member of the Omaha Tribe of Nebraska, is an anthropologist. He was appointed by the tribe to assist in the scientific study of Omaha cemeteries and Indian sites. He claims that "if the bones were reburied before they were adequately studied, more might be lost than gained for the tribe." His view is between the extreme all-or-none view held by many Indians and by many anthropologists and collectors of these relics and artifacts.

Your committee has been advised by a lawyer that most items in the collections of the city museums and university come under the 1990 law. Because there are no descendants of this extinct tribe today, there is no one to turn over these remains to for reburial. The lawyer advised the committee that it was free to do anything it wanted with the remains.

However, you are not free to decide when you will make a decision. At today's meeting, the mayor made a surprise appearance. She announced that a group of Native Americans representing ten different tribes will be visiting the community and university within the week, possibly even the next day. She does not want to have to debate the issue of the collections with these Indians. She wants a policy regarding these remains to be in force before the Indians arrive. She will follow any decision the committee makes, but that decision must be one and only one of the options she listed. These options are stated below.

1. Do nothing about the collection until 50,000 Native Americans visit the community and insist the remains be returned. Only when and if this happens will the relics and artifacts be returned. This return would occur within 1 year after the fifty-thousandth Native American makes his or her request. Half of all costs for this effort and transfer would be charged to the Native American tribes. All costs must be paid before any artifacts are moved.

2. Actively seek out one or more of the Native American tribes that have proof of being descendants of peoples whose remains and artifacts are in the museums. This search would mean directly contacting the leaders of every Native American tribe within 3 years and making arrangements for the transfer to these tribes. All costs for this effort and transfer would be charged to the Native American tribes.

3. Make a policy statement that all remains not falling under the 1990 law are the possession of the city and university. The remains will never be returned because of their scientific and public information value.

4. Direct the museum and university that their people will have 5 years to complete any scientific research using the Indian remains they have. At the end of the 5 years, all remains will be

reburied within 2 years. The conditions and location of this reburial will be decided upon with the assistance of Native Americans living in the state. Half of all costs for this effort and transfer would be charged to the Native American tribes. All costs must be paid before any artifacts are moved.

5. Direct the museum and university to donate all the remains and artifacts in their collections to a private collector who has agreed to build a single large museum and research laboratory on his own land. In this way the city and university will be rid of these collections but still have access to them. However, the private collector will own these remains and artifacts and will likely charge people who want to look at them.

6. Direct the museum and university to keep all artifacts and bones of Native American societies that have no present-day direct descendants. All materials from Native American peoples who have direct descendants today will be returned within 2 years. This would preserve about 60 percent of all the materials in the museums. All costs for this effort and transfer would be charged to the Native American tribes. All costs must be paid before any artifacts are moved.

You must make a decision, and you must make it soon. You are not free to make any decision you want. You must chose one and only one from the list the mayor has given you. To make the best decision possible, you need to do the following:

- Answer the questions on Predecision Task Sheet 1. These questions will help you review the story and important details about the laws regarding the artifacts.
- Complete Predecision Task Sheet 2. This task will help you think about each of the options you have available.
- Complete the Individual Decision Sheet.
- With members of a small group, complete the Group Decision Sheet, which gives the decision of your entire group.

PREDECISION TASK SHEET 1

This task will help you review and pay attention to important details in the story. Your answers to these questions will help you comprehend the story and the laws regarding Native American artifacts.

1. What are at least three reasons why a city or university would collect and display bones and artifacts of people from other cultures?
   a.

   b.

   c.

2. In determining the future of these bones and artifacts, what three roles might scientists play?
   a.

   b.

   c.

3. In locating, studying, and preserving the bones and artifacts, what are at least three areas of technology that might be used?
   a.

   b.

   c.

4. In your own words, what are all of the requirements of the Native American Grave Protection and Repatriation Act of 1990?
   a.

   b.

   c.

5. What are three beliefs many Native Americans have about Indian remains and artifacts that are kept in museums and universities?
   a.
   b.
   c.

6. What are three beliefs many scientists have about keeping Indian remains and artifacts in museums and universities?
   a.
   b.
   c.

7. Pretend you were a student who would be visiting these museums. If the bones and artifacts were removed, how would this affect your education about Native Americans of the past?

After you have completed your individual answers, share your responses with the others in your group. Add to, revise, or correct your responses as needed. Make sure everyone in your group comprehends this information before going on to the next page.

PREDECISION TASK SHEET 2

In the two columns, write the reasons you do accept or do not accept each of the six options you have available.

Choice	Reasons to Select This Choice	Reasons to Reject This Choice
1. Do nothing and wait until the fifty-thousandth Native American visits the museum.	a. b.	a. b.
2. Actively seek one or more tribes whose people will accept the remains.	a. b.	a. b.
3. Announce that the remains are the possession of the city and university.	a. b.	a. b.
4. Give the museum and university 5 years before returning the remains.	a. b.	a. b.
5. Give the collection to a private citizen.	a. b.	a. b.
6. Give back only the part of the collection that is linked to known direct ancestors	a. b.	a. b.

After you have completed your individual answers, share your responses with the others in your group. Add to, revise, or correct your responses as needed. Make sure everyone in your group comprehends this information before going on to the next page.

INDIVIDUAL DECISION SHEET

1. After considering these six options, the *best option* to select is

2. Three reasons why this is the best option are
   a.

   b.

   c.

3. The reaction of Native Americans to this decision is likely to be

4. The reactions of scientists, especially anthropologists, to this decision are likely to be

5. As a result of this decision, the city and university will *gain* the following four things:
   a.

   b.

   c.

   d.

6. Because of this decision, the city and university will likely *lose* the following three things:
   a.

   b.

   c.

7. In this situation, the *very worst* thing I believe we could do is to

GROUP DECISION SHEET

1. After considering these six options, as a group we believe the *best option* to select is

2. The three most important reasons why this is the best option are
   a.

   b.

   c.

3. The reaction of Native Americans to this decision is likely to be

4. The reactions of scientists, especially anthropologists, to this decision are likely to be

5. As a result of this decision, the four most important things the city and university will *gain* are
   a.

   b.

   c.

   d.

6. As a result of this decision. the three most important things the city and university will *lose* are
   a.

   b.

   c.

7. In this situation, the *very worst* thing we could do is to

REVIEW AND REFLECTION QUESTIONS

Suggested follow-up questions to focus and guide inquiry and learning.

a. What is an *anthropologist?* An *archaeologist?*

b. What is an *artifact?*

c. How might a scientist study the *remains* of past people?

d. By analyzing their bones, what three things might a scientist discover about a people?

e. If you were a Native American, what would your feelings be toward those who removed the bones of your ancestors from their burial sites?

f. If you were a Native American, what would you do to recover the remains of your ancestors from museums and universities?

g. If you knew for sure someone will unearth your bones and study them a hundred years after you are buried, what feelings would you have toward that person?

h. What is the main theme of this story?

i. What are the major characteristics of the race known as the Native American Indian?

j. What is a *race?*

k. What is a *culture?*

l. Would bones be needed to study a person's race or culture?

m. Suppose you were a Native American. In this story, what decision would you hope the committee made?

n. Suppose you were a Native American. In this story, what would be the worst decision the committee could make?

o. Suppose you were an anthropologist. What would be your attitude toward Native Americans who wanted these remains returned for reburial?

p. Suppose anthropologists started to excavate and study the remains of early white or Hispanic settlers. What would your feelings be toward these scientists?

q. Suppose anthropologists started to excavate and study the remains of your parents or grandparents. What would your feelings be toward these scientists?

r. Under what conditions should anthropologists *be allowed to* unearth and study ancient remains?

s. Under what conditions should anthropologists *be prevented from* unearthing and studying ancient remains?

t. If you were in this situation and could make *any* decision, what would your decision have been?

u. In what ways is the work of an archaeologist different from the work of an anthropologist? In what ways are they identical?

v. What rights do people have to dig up, study, and display the bones and artifacts of people of earlier days?

w. Suppose someone told you they were going to dig up your body after your were buried. Suppose he or she also told you your bones would be studied and then put on display in a museum. What would be your reaction to these plans?

x. Suppose it was against the religious beliefs of a society for the bones and artifacts of the dead to be dug up once buried. To what extent should people from cultures who did not believe the same thing be able to dig up the dead of this society?

y. Suppose you had a chance to visit a museum such as those described in this story. What would be your reasons for visiting or not visiting the displays of Indian bones and artifacts?

z. Suppose people today could not study the remains and artifacts of humans who lived long ago. What effects would this have on our knowledge of the Earth's past?

CHOOSE OR LOSE

The end of the year is fast approaching. The manager of your city announced there would be a sizable sum of money left over at the end of the year. According to state law, all unspent moneys must be returned to the state treasury.

You are a member of the city council. You have waited a long time for extra money to be available to spend. For years you have sought support for three of your pet projects. In fact, just last year the city council voted to spend any extra money for your projects. However, there was not enough money left in the treasury to pay for the projects. You have wanted the money to

1. build bike, running, and walking paths
2. build and operate a city park and small animal zoo
3. purchase a large piece of land to be used as a natural wildlife refuge

CHOOSE OR LOOSE

Now it appears your three projects will be funded.

This evening the city council is meeting for the last time this year. The decision to spend the money must be made tonight. You have already reminded the council members of their vote 12 months ago to fund your three projects when money was available.

As the city manager begins the meeting, your dreams are shattered. Only one-third of the money that was thought to be available is still unspent. Of your three projects, only one can be funded. The council members inform you they will spend the money on one of your projects. It is your choice. However, they cannot promise or guarantee extra money will be available next year or the next. To delay your decision would force them to fund other projects that are proposed by other members of the council.

In other words, if you decide not to make a choice, you will lose all of your projects. When you decide on one of the three, you will probably never get your other two projects funded. The council members agree you must choose from the three projects you have long supported. You must choose one and only one of the three projects.

You are to decide which of these projects you most want to be funded. You then must convince other members of the committee to select the best project to fund. To help you consider these three options before you make your decision, complete the Predecision Task Sheet on the next page by yourself.

Only after you have completed the Predecision Task Sheet are you to make your personal decision. Record your personal decision on the Individual Decision Sheet.

Suppose you were a member of a committee that was to make a group decision as to what option to support. After you have made you individual decision, work together as a member of a group to reach a consensus decision. Record your group's decision on the Group Decision Sheet.

PREDECISION TASK SHEET

This task sheet will help you consider important details about the story and the options. Write in responses to all of the areas and questions listed below *before* you make a final decision.

Projects	Ways This Might Protect the Environment	Ways This Might Harm the Environment	What Groups Would Benefit Most From This Project?
1. Bike, running, and walking paths	**a.**	**a.**	**a.**
	b.	**b.**	**b.**
2. Park and zoo	**a.**	**a.**	**a.**
	b.	**b.**	**b.**
3. Wildlife refuge	**a.**	**a.**	**a.**
	b.	**b.**	**b.**

1. Which one of these projects would environmentalists most likely support?

2. What would be three important reasons why environmentalists would support this project?

   a.

   b.

   c.

3. Of these three projects, which one would most scientists most likely support?

4. What are three important reasons why most scientists would probably support this project?
   a.

   b.

   c.

5. Of these three choices, the one I would most use would be

6. Two of the strongest reasons why I would use this one the most are
   a.

   b.

After you have completed your individual answers, share your responses with the others in your group. Add to, revise, or correct your responses as needed. Make sure everyone in your group comprehends this information before going on to the next page.

INDIVIDUAL DECISION SHEET

On this sheet, record your personal decisions regarding which of the three options you will vote for funding by the city council. The other two projects will not be funded. The three projects that could be funded by the council are

- Build the bike, running, and walking paths.
- Build and operate a city park and small animal zoo.
- Purchase land to be used for a natural wildlife refuge.

1. As a member of the city council, my attitude about the environment should be

2. I believe that in the area of the environment and spending public funds, I should be especially concerned about these three things:
   a.
   b.
   c.

3. Of these three projects, the best project is

4. The three strongest reasons why this decision is the best are
   a.
   b.
   c.

5. My decision in this situation will cause people in this city to lose the following two programs:
   a.
   b.

Now that you have made your individual decision, work with your group to arrive at a consensus decision. You do not vote. You are to arrive at a group decision all members of your group will accept. Record your group decision on the Group Decision Sheet.

GROUP DECISION SHEET

Members of your group are to agree on one and only one of these three projects. You should have the same reasons for agreeing to this one choice. This means you are not to vote. You must reach a common conclusion all members of your group willingly accept and support. The three projects that could be funded by the council are

- Build the bike, running, and walking paths.
- Build and operate a city park and small animal zoo.
- Purchase land to be used for a natural wildlife refuge.

1. As members of the city council, our attitude about the environment should be

2. We believe that in the area of the environment and spending public funds, we should be especially concerned about the following three things:

a.

b.

c.

3. Of these three projects, the best project is

4. The three strongest reasons why this decision is the best are

a.

b.

c.

5. The option we would most want to see not funded is

The persons responsible for making this decision are

________________________ ________________________

________________________ ________________________

REVIEW AND REFLECTION QUESTIONS

Suggested follow-up questions to focus and guide inquiry and learning.

a. Remembering that bicycle paths might lead some motorists to stop driving to work, which of the three projects would be the most protective of the environment?

b. In what ways are these projects designed to conserve or preserve the environment?

c. Which project would reflect the *wisest* use of the available money?

d. Is it bad that projects such as these were not already funded by the city council?

e. What projects that operate to preserve and protect the environment are supported by your own city's government?

f. The building of roads and parking lots has been condemned by many environmentalists. Would an environmentalist who worked against the building of roads support the bike path project?

g. What does it mean to *conserve* the environment?

h. What does it mean to *preserve* the environment?

i. What is the major difference between *conserving* and *preserving* the environment?

j. What could scientists do to help *conserve* the environment?

k. What could scientists do to help *preserve* the environment?

l. Suppose one of the members of the city council were a scientist. What kinds of help would you expect that person to provide to the council?

m. Suppose the city council said you could hire a scientist to get information to help you make your decision. What kinds of information would you want this scientist to get you?

n. What kinds of information could scientists give to members of the city council in your city?

o. What kinds of scientific information could you give to members of the city council in your city?

p. Of these three projects, which one would probably require the most scientific information to be successful?

q. What kinds of scientific information would be needed to complete each of these projects?

r. If you were a member of a group that had to make a decision such as this, where would you go to find information that would help you make this decision?

s. What would be three ideas most environmentalists would likely accept?

t. Which of the three options in the story might harm the environment?

u. What specific aspects of technology would be needed for each project?

v. Which project would require the least technology? The most technology?

w. What connections exist between the technology that will be used for each project and the amount of harm that may be done to the environment?

TOO FEW OF A GOOD THING

You are chairperson of the annual, week-long Winter Festival in your community. You have made arrangements for the most popular rock group in the country to appear on the opening night of the festival. You seem assured of a successful evening because thousands of ticket inquiries are pouring in from a three-state area.

Unfortunately, the only suitable facility for the performance is the Civic Auditorium, which will only accommodate 5,000 people. Due to the weather, all festival activities must be held inside. Hence, no outdoor concert is possible.You have a seating problem.

You meet with the planning committee to discuss the problem of the high demand for a scarce number of seats. Elizabeth Wall, a local business leader, speaks about ways economists view the issue of scarcity. She asks that the committee think about this information as it seeks to make its decision. Wall explains that a basic economic problem all societies face is scarcity. *Scarcity* means there is less of something than what is needed or wanted. A *scarce resource* can be goods, services, raw materials, money, time, or any item of which there is not enough to satisfy a particular want. We all have unlimited wants, but we are faced with limited resources to satisfy them.

In this situation, we have a scarcity problem because the demand for tickets is greater than the 5,000 seats available. If only 5,000 people wanted tickets, we would have no scarcity problem.

Every society faces the problem of how to distribute scarce items. In fact, the central focus of economics is how a society or group (a) *allocates scarce resources,* (b) *produces products and services,* (c) *distributes them to its members,* and (d) *consumes them.*

We must decide how to allocate the 5,000 tickets to those who want seats. There are many mechanisms a society or group can use to distribute scarce resources. Sometimes the methods are created through careful planning. In other cases, the methods just evolve. Among the ways to distribute scarce items are

- *Brute force.* This method focuses on use of or risk of bodily harm to determine who will get the resource. Using this method, the committee could set up a situation where people who wanted tickets would be forced to physically fight to determine who would get them.
- *Queuing.* This method focuses on having people line up with distribution on a first-come, first-served basis. This method is used for making purchases at gas stations or ice cream parlors. Using this method, the committee could open ticket booths and sell tickets to persons in line. Or the committee might ask for mail-in requests. In both cases, discussions would be on a first-come, first-served basis.
- *Random selection.* This method focuses on pure chance using such games as a lottery, throwing dice, and drawing names or numbers from a box. Using this method, the committee could hold a lottery, raffle, or drawing to determine the 5,000 ticket "winners." The "losers" would get their money back.
- *Government action.* This method focuses on having state or local government make the decision. Such decisions are expected to be based on equal shares for everyone or on personal needs. Since more than 5,000 people "need" tickets, it is not clear how the government would decide who *most* needs them. Nor is it clear how the government would share 5,000 tickets "equally" among all those who want them. Using this method, the committee would turn the decision over to the government.

- *Marketplace/Auction.* This method focuses on the use of formal or informal arrangements between buyers and seller about the prices to be set for goods. These arrangements result in agreements to trade goods or services for money or other items. Depending upon supply and demand, the price for goods could rise or drop. Using this method, the committee could auction off the tickets to the highest bidders because the demand certainly is far greater than the supply.

After thinking about the different methods people use to deal with scarcity, the committee has agreed to consider only five options. All other options are only a little different from these five.

1. Have all persons who desire tickets meet on an open field where the tickets will be dropped from a plane. All persons who are strong enough to get and keep tickets in the struggle will be able to pay for the tickets and attend the concert.
2. Announce a time when tickets will be sold on a first-come, first-served basis. Only the first 5,000 people at the booth will be able to buy tickets. Costs for the tickets will be reasonable.
3. Have individuals mail in money for tickets. At a given date, all envelopes will be put into a box. Only the winners of the drawing will get tickets.
4. Turn the entire issue of ticket sales over to the city council for them to decide who gets tickets and who doesn't on the basis of equality or need.
5. Have a public auction with the tickets going to the highest bidders. Only those who have enough money and desire to outbid others will get tickets.

The decision of the committee is very important in that

- Opening-night attendance sets the stage for the ultimate success of the Winter Festival.
- The Festival is an annual event. As chairperson, you must not make a decision that will offend future festivalgoers.
- The Winter Festival is your community's most important money-making project for the local children's hospital.

You must move toward making a decision as a committee. Time is a problem. You must have time to carry out your decision once made.

Before you make your group decision, review the situation to make sure you comprehend the situation and the options. To help you make the best possible decision, complete Predecision Task Sheets 1, 2, and 3 before going on to the Decision Sheets.

PREDECISION TASK SHEET 1

1. In your own words, what does *scarcity* mean to economists?

2. What does it mean for something to be *scarce?*

3. How does *supply and demand* determine the scarcity of an item?

From the story and your own thoughts, take time to paraphrase and then review the five methods of dealing with scarce goods. The chart will help sort out your thoughts.

Way to Distribute Scarce Things	Possible Positive Consequences	Possible Negative Consequences
1. Brute force	a.	a.
	b.	b.
2. Queuing	a.	a.
	b.	b.
3. Random selection	a.	a.
	b.	b.
4. Government action	a.	a.
	b.	b.
5. Marketplace/Auction	a.	a.
	b.	b.

Before going to the next page, share your answers with others in your group. Add to your list the ideas your peers suggest. Make sure all members of your group comprehend this information and the story before moving to Predecision Task Sheet 2.

PREDECISION TASK SHEET 2

Complete this Task Sheet alone. Consider the issues of scarcity, wants, and distribution of scarce resources in light of critical values within a democratic society. The questions below will help you consider these values prior to making your final decisions.

1. What does it mean to be "fair"?

2. If you wanted to be fair, what option(s) would you choose?

3. What does it mean to be "equitable"?

4. If you wanted to be equitable, what option(s) would you choose?

5. What does it mean to be "ethical"?

6. If you wanted to be ethical, what option(s) would you choose?

7. Which option(s) would be the most "unfair"?

8. Which option(s) would be the most "inequitable"?

9. Which option(s) would be the most "unethical"?

Before going to the next page, share your answers with others in your group. Add to your list the ideas your peers suggest. Make sure all members of your group comprehend this information and the story before moving to Predecision Task Sheet 3.

PREDECISION TASK SHEET 3

Complete this Task Sheet alone. From information in the story, the committee has to select from only five choices. Use the chart below to help you review each choice.

Choice	Possible Good Reasons for Using This Method	Possible Bad Reasons for Using This Method
1. Tickets dropped from a plane—have people fight for them	**a.** **b.**	**a.** **b.**
2 Tickets sold on first-come, first-served basis at a ticket booth	**a.** **b.**	**a.** **b.**
3. Tickets drawn from a box after people have mailed in requests	**a.** **b.**	**a.** **b.**
4. Tickets distributed by the city government on basis of equality or need	**a.** **b.**	**a.** **b.**
5. Tickets sold at public auction to the highest bidders	**a.** **b.**	**a.** **b.**

After you have completed your individual answers, share your responses with the others in your group. Add to, revise, or correct your responses as needed. Make sure everyone in your group comprehends this information before going on to the next page.

INDIVIDUAL DECISION SHEET

It is time to make your final choice. Your committee has agreed that only one of the options can be selected. In the space below, mark an *X* to the left of the option you prefer most.

_____ **1.** Drop the tickets into a crowd and let them fight.

_____ **2.** Make people stand in line on first-come, first-served basis.

_____ **3.** Have a drawing based on chance.

_____ **4.** Turn the issue over to the city government for their decision.

_____ **5.** Have an auction for the highest bidders to get tickets.

1. The two most important reasons why I like this option best are

a.

b.

2. The two best things that will happen with this choice are

a.

b.

3. What worries me most about making this decision is

4. The choice I disliked the most was option ______ for the following two reasons:

a.

b.

GROUP DECISION SHEET

It is time for your group to make its final choice. Your committee has agreed that only one of the options can be selected. You are to make a consensus decision rather than deciding by majority rule. In the space below, mark an *X* to the left of the option the entire group most prefers.

_____ **1.** Drop the tickets into a crowd and let them fight.

_____ **2.** Make people stand in line on first-come, first-served basis.

_____ **3.** Have a drawing based onchance.

_____ **4.** Turn the issue over to the city government for their decision.

_____ **5.** Have an auction for the highest bidders to get tickets.

1. The two most important reasons why we like this option best are

a.

b.

2. The two best things that will happen with this choice are

a.

b.

3. What worries us most about making this decision is

4. The choice we disliked the most was option ______ for the following two reasons:

a.

b.

REVIEW AND REFLECTION QUESTIONS

Suggested follow-up questions to focus and guide inquiry and learning.

a. What does the word *economics* mean to you?

b. What are the names of five things you know that are "scarce"?

c. What makes these things scarce?

d. What do you own that someone else might label as scarce?

e. People who study the economy and the buying and selling of things are often referred to as scientists. What are at least three reasons why they would be scientists?

f. For what reasons would scientists study the economy of a country?

g. What sorts of things would a scientist probably study about the economy?

h. How might a scientist go about investigating and doing research on the economy?

i. What tools or equipment would an economic scientist probably use to collect and analyze data about the economy of a nation?

j. How might the scientific study of the economy help people in a society? How might it help you?

k. In the story, how could the committee have made tickets to the event less scarce?

l. What is the opposite of being scarce?

m. What kinds of problems does scarcity create?

n. Why should people study things such as scarcity and economics?

o. Why is it good that people such as you study problems of scarce things?

p. What was the most important idea about economics that you picked up from this episode?

q. How is the price of an item related to the need for that item?

r. In this story, if you wanted to buy a ticket for the show, which option would you want the committee to adopt?

s. Of the options in the story, which option was the most dangerous?

t. Of the options listed in the story, which option was the most fair?

u. If certain items could be produced in larger numbers but aren't, to what extent is the speed of production making that item scarce?

v. If items people need become too scarce, what possible actions might they take in reaction to this scarcity?

w. How might technology be used to implement each of the five options?

x. In your city, how is technology used to sell tickets to popular events?

y. What is the major difference between a want and a need?

z. In what ways can we prevent something from being scarce?

EPISODE 5

THE CIRCLE

Each day since the car accident Anna's life has become more hectic. Her once-favorite room, the den, resembles a hospital room. Her younger sister, Alison, has been home from the hospital for about six months. Alison is still semiconscious.

Anna and Alison had always been close. They shared many secrets. Anna hoped that some day she would be able to play with Alison again. The doctors were honest with Anna. They told her they thought Alison would live. They could not tell her whether Alison would ever walk or talk again.

Their mother finally accepted that Alison had permanent brain damage. She convinced doctors that Alison could be cared for as well at home as at the overcrowded hospital. The doctors admitted that the familiarity of her home might trigger more responses from Alison.

A hospice home-care team was set up in Alison's home. The members of the team were Alison's doctor, a registered nurse, and a physical therapist. The nurse spent each morning with Alison, and could be reached by phone at any time. The physical therapist worked with Alison every other afternoon. During the first few weeks, Alison's doctor visited daily. He prescribed her medicine and communicated with the hospice team. Each of these people showed Anna and her mother how to take better care of Alison.

Alison had remained the same during the last month. She focused her eyes on various objects and people. School friends made tapes for her to listen to and drew pictures to decorate her room. The doctors explained that the more Alison heard and saw familiar things, the sooner other parts of her brain might begin to function. Friends of her mother often stopped by to rock Alison or to help feed her. Alison appeared to understand more than she could verbalize. Sometimes she nodded her head when asked a question; sometimes the expression on her face changed. She even seemed to smile.

One evening, Anna noticed that her mother was very quiet. Her mother had just been told Alison was going to die. The doctor had told her it would be best for Alison and Anna to know.

The doctor had described the stages a family went through after a death. He said it was better for Anna to know the facts. If the family didn't prepare now, the grieving would be worse after Alison died.

Anna's mother remembered how, when the accident first happened, she talked to Anna about the possibility that Alison might die. Back then, she did not really think Alison would get worse, much less die. Now she knew Alison would die; she just was not certain when.

As Alison's and Anna's mother, she must decide whether she will tell Anna and Alison. She must also decide when and how to tell them. As she considers her decision, she realizes she has few choices open to her. These are the things she could do:

1. Never tell Anna or Alison.
2. Tell Anna and Alison immediately.
3. Tell Anna and not Alison.
4. Tell Alison and not Anna.
5. Have the doctor tell Anna and Alison.

The mother knows these are the only real choices she can make. If she waits too long, Alison could die. She also knows that whatever her choice, she must make it now.

Imagine you are the mother of Anna and Alison. Imagine the decision is yours to make. Read and study the Information Sheet that follows. Then complete the Predecision Task Sheet before making your decision about whether to tell Anna and Alison.

INFORMATION SHEET

A psychologist named Elisabeth Kübler-Ross studied how people think about their own death. She used the observation and record-keeping skills of a scientist to collect information. She listened to dying people talk to their families, doctors, and ministers. She took careful notes on their actions and conversation. After spending many months analyzing and interpreting the data she had collected, she and her associates came up with new ideas about how people react toward their death. Her ideas have changed our view about how to deal with those who know they are going to die.

According to Dr. Kübler-Ross, most people who know they are going to die go through five stages (or levels) of thinking.

1. The first stage is *denial.* The person does not accept the fact that he or she will die. Typically the person rejects the news as not being true or as being a joke. When the sickness becomes so bad the person cannot ignore the fact any longer, the person usually moves to the next stage.

2. The second stage is *anger.* The person gets upset and angry at almost everyone. Part of the reason for this is that the person is upset that he or she, not someone else, is going to die.

3. The third stage is *bargaining.* The person tries to delay death by asking for more time to do good things instead of having to die. The person may promise to go to church, give more to charity, or do other good things if she or he can be allowed to live longer.

4. The fourth stage is *increased depression.* The person finally realizes he or she is going to die and that there is nothing that can be done to prevent it. The person may withdraw from others and spend time grieving for and about their own death.

5. The fifth stage is *acceptance.* Here the person accepts the fact that he or she will die. The person is relaxed and at peace with himself or herself.

Dr. Kübler-Ross calls these five stages a *circle of suffering.* These are not required ways of thinking. If people are not allowed to think and act in these ways in this order, their suffering lasts longer and is more painful. She says that if you want to help a dying person, don't try to cheer them up. Instead, help them with their thoughts on each step when they are at that particular step. Many people never get to the fifth stage, but remain at one or more of the earlier stages. Meanwhile, some people move through to the fifth stage very rapidly. Keep in mind that if a person does not know she or he will die, these steps do not occur.

Either as an individual or a member of a small group, paraphrase each of the five stages. Write brief answers to the questions below.

1. What are the names of the five stages in their correct order?

Stage 1:

Stage 2:

Stage 3:

Stage 4:

Stage 5:

2. In your own words, what does a person think about in each of the five stages?

Stage 1:

Stage 2:

Stage 3:

Stage 4:

Stage 5:

3. When these stages are interrupted or prevented from taking place, what is likely to happen to the person's suffering?

4. What are at least three ways you might use this information on the circle of suffering?

a.

b.

c.

Before going on to the Predecision Task Sheet, review your responses with one or more of the students in your class. Add to or revise responses to make them more complete and accurate.

PREDECISION TASK SHEET

Suppose you were Alison's mother and trying to decide the best thing to do. The five choices below are the only choices you have at this time. You cannot delay your decision any longer. To help you make your decision, fill in the chart.

Choice	Reasons Why This Should Be Done	Reasons Why This Should Not Be Done
1. Never tell Anna or Alison	**a.** **b.**	**a.** **b.**
2. Tell Anna and Alison immediately	**a.** **b.**	**a.** **b.**
3. Tell Anna and not Alison	**a.** **b.**	**a.** **b.**
4. Tell Alison and not Anna	**a.** **b.**	**a.** **b.**
5. Have the doctor tell Anna and Alison	**a.** **b.**	**a.** **b.**

Before going on to the Decision Sheet, share your reasons for each of these choices with one or more of the students in your group.

DECISION SHEET

Complete this section by yourself.

1. What is death?

2. What does it mean when someone dies?

3. When someone dies, what happens to that person?

4. Of the information about the *circle of suffering,* what information should Alison's mother use in making her decision?

5. If you were Alison, what decision would you want your mother to make?

6. After considering the choices, if you were Alison's mother, what decision would you make?

7. The three most important reasons for this decision are
   a.

   b.

   c.

8. This decision is likely to affect Alison in the following ways:
   a.

   b.

   c.

9. This decision is likely to affect Anna in the following ways:
   a.

   b.

   c.

10. This decision is likely to affect Alison's mother in the following ways:
   a.

   b.

   c.

Once you have completed these responses by yourself, share your responses with one or more of the students in your group. Listen carefully to what they say and the reasons they made their choices.

GROUP DECISION SHEET (OPTIONAL)

For this decision task, work together to develop a single set of responses to the questions and statements below. Make your group decision by consensus. This means your group can only write answers in areas where everyone in the group fully agrees with the response.

1. What is death?

2. What does it mean when someone dies?

3. When someone dies, what happens to that person?

4. Of the information about the *circle of suffering,* what specific information should Alison's mother use in making her decision?

5. If you were Alison, what decision would you want your mother to make?

6. After considering the choices, if you were Alison's mother, what decision would you make?

7. The three most important reasons for this decision are
   a.

   b.

   c.

8. This decision is likely to affect Alison in the following ways:
   a.

   b.

   c.

9. This decision is likely to affect Anna in the following ways:
   a.

   b.

   c.

10. This decision is likely to affect Alison's mother in the following ways:
   a.

   b.

   c.

RESEARCH ON DYING DECISION SHEET (OPTIONAL)

Death is a social affair. Individuals who are dying do so in a social environment. Others in society may be affected by the dying or death of an individual before, during, and after his or her death. Until the late 1960s, little research was done on death and dying. Now there is even research on how people react to and try to adjust to the death and dying of others. Answer the questions below to explore issues that tie research to concerns about death and dying.

1. What are at least three ways society might be affected by the death of a person?
   a.

   b.

   c.

2. From what you read about Dr. Kübler-Ross' findings, in what ways would her work be considered
   a. scientific?

   b. not scientific?

3. What would be at least three benefits to society of Dr. Kübler-Ross' research findings?
   a.

   b.

   c.
4. What are at least two ways her research findings might be relevant to your own life?
   a.

   b.

5. Suppose someone you knew was dying. How might you use this information to help him or her complete the *circle of suffering?*

REVIEW AND REFLECTION QUESTIONS

Suggested follow-up questions to focus and guide inquiry and learning.

a. From the information given on the Information Sheet, in what ways is Dr. Elizabeth Kübler-Ross a scientist?

b. Why would a scientist study the way people think about their own death?

c. What is a psychologist?

d. What are some other areas related to dying that scientists study?

e. To what extent might it be *good* for people who are not dying to talk about death?

f. In what ways might it be *bad* for people who are not dying to talk about death?

g. In what ways might the information about the circle of suffering be helpful to people who are dying?

h. In what ways is this information about the circle of suffering likely to be helpful to people who are not dying?

i. Now that you have this information about the circle of suffering, how might you use it?

j. If you did not use this information when you could use it, would you be doing the best thing you could do in that situation?

k. In what ways would the information about the circle of suffering be scientific information?

l. What makes a piece of information "scientific"?

m. In what ways might the information on the *circle of suffering* be helpful to a person who knows someone who is dying?

n. When someone dies, what happens to their body?

o. When someone dies, what happens to their personality?

p. When someone dies, what happens to that person?

q. To what extent does a person live in another's mind rather than their own body?

r. Suppose a favorite pet of yours died. How might the five stages of the circle of suffering be helpful to you in adjusting to this loss?

s. On a scale of 1 to 10, with 10 representing a great contribution to science, how would you rate Dr. Kübler-Ross' discovery of the *circle of suffering?*

NO DEEP BREATHS

Jim Weyand has been working at the textile mill in Centerville for 3 years. He had considered himself lucky to have a job. There just weren't any better jobs for unskilled or skilled laborers in this small town. The mill paid only minimum wage for unskilled laborers. Skilled workers such as Jim Weyand were not paid much more—the maximum salary was just over twice the minimum wage. Very few workers ever worked long enough to get this high salary.

Many of the workers at the mill respect Jim. They recently talked to him about their working conditions. The constant loud noise from the machinery during the 8-hour workday had caused temporary deafness in nearly all the mill workers. For some the deafness was permanent. There were few breaks for rest and relaxation, and the workers hated the monotony of their routine tasks. Because of low wages and job insecurity, they had even stopped asking for the restrooms and cafeteria to be cleaned more often.

One of the worst things about the job was the amount of dust and fine dirt filling the air. It constantly bothered the workers, who had to breath it all day. The signs on the walls warned workers "Don't Breathe Deeply" and "Don't Take Any Deep Breaths."

Workers coughed almost all the time. The poor ventilation did not help bring fresh air into the mill. The workers could not escape the dust-filled air. One worker remarked, "Inside this building it is like breathing smog."

Last year, Leo Lapre died. His doctor said he died from brown-lung disease, caused by 20 years of working in the mill dust. Leo was only 42 years old.

Brown-lung disease is caused by inhaling dust that then gets trapped in the lungs. The dust clogs the lungs and makes breathing difficult. The person begins to get short of breath and to cough to clear the infected lung. The diseased lung caused by cigarette smoke is one example of brown lung. Eventually a person may be unable to breathe. Many people working in mills, coal mines, and dust-filled smoggy cities die of brown-lung disease each year. Leo had asked Jim to do something to rid the mill of dust.

Over the past 3 years, a number of scientists studying the health of employees had expressed concern for the health of the workers at the mill. They found the average life expectancy of workers at the mile was almost 20 years below that of people in other occupations in the same community. The average level of health of workers was far below the national average. Meanwhile the incidences of lung-related diseases and infections was seven times the national average. The scientists' reports were welcomed by the workers because someone was finally being done.

Government and university researchers reported on a number of ways to cut the air pollution in the mill. They claimed that the technology was available to significantly reduce the pollutants in the air. However, the owners would need to spend over $100,000 in the first 2 years. In addition, the owners would need to change several of the policies regulating mill operations.

The owners and managers of the mill ignored the scientists' health recommendations. Being the largest employer in the area, they were confident the workers and their families would not do anything against the mill. Besides, the mill owners said the scientists they had hired did not find the health problems the government researchers had found.

Jim had read these reports. He knew Leo's death was part of the statistics the scientists had announced. Jim was also aware of the scientists who had studied the economy of the region. It was clear from the economic data that without the textile mill and the salaries of the workers, the area would be very poor. Most people would be unemployed. The quality of the local school, public programs, and health facilities would suffer. Jim knew the two sets of data made a clear-cut answer difficult for the mill workers.

Several of Jim's friends wanted him to be their leader in getting the managers of the mill to listen to them. They all realized they could not afford to be without their salaries for a long period of time. Yet, how long could they continue to work in the existing conditions?

Jim agreed to find ways to show the managers that the workers needed better working conditions. He called his brother, who was a member of a labor union at a larger textile mill in another town. His brother outlined the following possibilities:

1. *Workers could boycott the company's products.* This means you, your family, and your friends will not buy any products made by the company until your demands for better working conditions have been met. You could even try to get people you do not know to stop buying the company's products.
2. *Workers could go on strike.* This means you not only stop working, but openly try to persuade others to stop as well. You might even have to force others to stop working for the company. However, once on strike you lose your regular take-home pay and any hope for a bonus or health benefits.
3. *Workers could resign.* This means you no longer report to work and must look for another job. This would have to be an individual decision. If enough of you resigned, it might take several weeks to replace you. This would cost the mill a lot of money. However, there are no other employers in the town. The nearest town where a job is likely is over 100 miles away.
4. *Workers could commit acts of violence and sabotage against the company.* These activities are planned to damage and destroy the property of the company. They are criminal acts. If caught, you would be subject to fines and prison terms.
5. *Workers could participate in a work slowdown.* This means you stay on the job but reduce the amount of work you do so the company does not produce as much. This reduces the quantity of production. If caught, individuals are usually fired and never rehired.

Jim told the other workers about what his brother said. They agreed something had to be done. When Jim told his bosses about what the workers said, they ignored him. The bosses refused to spend money to clean the mill dust from the air. They would not get air filters or conditioners. They wouldn't even spend money to clean up other areas around the mill. Even worse, they would spend no money to help pay the medical bills of the workers. Workers with brown-lung disease would receive no special treatment.

The answers from the mill bosses upset Jim and his fellow workers. At the meeting, the workers agree to make a choice about what to do.

Imagine that you are Jim or one of his coworkers. In this situation, what will you do?

Before you go on with the story, take a few minutes to review the information about Jim's work and the conditions at the mill. Share your ideas about the details of the story with members of your group. Write answers to the questions on Predecision Task Sheet 1 before moving to the second Predecision Task Sheet. Make sure they all comprehend the story.

PREDECISION TASK SHEET 1

These questions will help you comprehend important information in the story. Take time to write complete answers.

1. What is brown-lung disease?

2. What are the causes of brown-lung disease?

3. What is the major theme of this story?

4. In what ways might the use of technology lead to people getting brown-lung disease?

5. In what ways might the use of science lead to people getting brown-lung disease?

6. In what ways might the use of technology prevent people from getting brown-lung disease?

7. What are at least three reasons why workers should not be forced to work under conditions that might lead to brown-lung disease?
   a.

   b.

   c.

8. What are at least three reasons why a company might continue to provide working conditions that contribute to brown-lung disease?
   a.

   b.

   c.

After you have completed your answers, share your responses with the others in your group. Add to, revise, or correct your responses as needed. Make sure everyone in your group comprehends this information before going on to the next page.

PREDECISION TASK SHEET 2

Suppose you were among the friends to whom Jim was talking. The five choices below are the only choices you have at this time. As an individual or member of a group, fill in the chart. This will help you think about the reasons for each choice before you make your final decisions.

Workers' Choice	This Would Be the Best Thing to Do Because . . .	This Would Be the Worst Thing to Do Because . . .
1. Begin boycotting policies	a.	a.
	b.	b.
2. Go on strike	a.	a.
	b.	b.
3. Resign from the mill	a.	a.
	b.	b.
4. Commit acts of violence	a.	a.
	b.	b.
5. Participate in a work slowdown	a.	a.
	b.	b.

After you have completed your answers, share them with the others in your group. Add to, revise, or correct your responses as needed. Make sure everyone in your group comprehends this information before going on to the next page.

INDIVIDUAL DECISION SHEET

Individually you must decide which of the five choices will work best for you. Once you participate in any one of these activities, you may be fired. You want to make the best choice to get the mill managers to listen to you and improve working conditions.

1. I choose number ________ as my most preferred choice because
   a.

   b.

2. Number ________ is unacceptable to me because
   a.

   b.

3. Number ________ is unacceptable to me because
   a.

   b.

4. Number ________ is unacceptable to me because
   a.

   b.

5. Number ________ is unacceptable to me because
   a.

   b.

6. The option I selected as my most preferred choice will produce the following results:
   a.

   b.

   c.

7. The most frustrating part about this situation is

GROUP DECISION SHEET

You are a member of a team of workers who has agreed to make one decision, which all members will follow. As a group you must decide which of the five choices will work best. Once you participate in any of these activities, you may be fired. You want to make the best choice to get the managers to listen and to improve working conditions.

1. Number ______ is our most preferred choice because

2. Number ______ is the most unacceptable to us because

3. The option we selected as our most preferred choice will produce the following results:

4. We hope the mill owners will use science in the following ways to improve working and health conditions in the mill:

5. To improve working and health conditions in the mill, we hope the mill owners will use technology in the following ways:

6. The very worst thing we could do is to

7. The very worst thing the mill owners could do is to

8. If asked to justify our decision to our families, we would say

REVIEW AND REFLECTION QUESTIONS

Suggested follow-up questions to focus and guide inquiry and learning.

a. According to the story, what is brown-lung disease? What is one cause of it?

b. In the story, a group of scientists collected data about the health of the people in the community. What kinds of information would they want to know about people's health?

c. In the story, a group of scientists collected data about the economic life of the community. What kinds of information would they gather to study the economy of a group of people?

d. For what reasons would scientists study the health of people in a mill?

e. What did these scientists do with their data about the health of the workers in the mill?

f. What should these scientists have done with their data?

g. How can the scientists who collected these data help workers who are endangered in their jobs?

h. What did the economic scientists do with their data about the economic life of the community?

i. To what extent were the data and conclusions of the health and economic scientists helpful to the mill workers? To the factory owners?

j. If these data and conclusions were not very helpful, why should scientists collect such data?

k. How might scientists have helped the workers in the mill?

l. How could the workers have used the information reported by the scientists?

m. What other kinds of information might scientists study about a workplace or community?

n. Who should pay the scientists to collect this information?

o. Suppose the mill owners had paid the scientists to study the workers' health. What does this story suggest about how people who pay for research pay attention to the results?

p. If you were Jim and you could do anything you wanted, what would you have done?

q. If you had acted in this way, what would have been the consequences for you?

r. What is the major theme of this story?

s. In this story, what role did technology play in the lives and health of the workers?

t. In this story, did science or technology have the biggest impact on the lives of the workers?

u. If you were a worker in this mill, what feelings would you have toward the mill owners for not improving your working conditions?

v. If you were a worker in this mill, what feelings would you have toward the government for letting these working conditions continue at the mill?

w. If you were one of the mill owners, what feelings would you have toward the scientists? The government? The workers?

CLIFF HANGER

You have been on several camping trips with your family. Each time, your family drove to the campsite. Then after a day of fun and play, you slept in a cabin. Sometimes, when friends came along, you slept in a tent. While you did a lot of walking around, you wanted to learn how to hike. Even more, you wanted to learn how to rappel down the short rocky cliffs of the nearby campsites.

Your older brother and sister taught you the secrets of hiking. You realized hiking was more than just long walking. They told you there are certain things you must do to be a good hiker. Among the things they told you were

- Always carry a first aid kit.
- Establish a good pace, and shorten steps when terrain gets rough.
- Look for and stay away from poisonous plants.
- Wear well-fitting, comfortable shoes and lightweight, loose-fitting clothes.

- Observe and pay careful attention to landmarks to help you know where you are and where you have been.
- Mark the trail well so you can find your way back—even when you are going for what you think is a short walk.
- Take a 10-minute break every hour.
- Step around, not over, obstacles.

Your scout troop had a rappelling badge you were anxious to acquire. Working very hard, you completed the requirements for the badge. To earn your badge, you learned the rules of safe rappelling and even had to rappel off a 20-foot-high cliff. Some of the rules you followed in your rappelling tests were:

- Estimate the length of rope needed to reach the bottom, and double the amount.
- Set up the rope for rappelling and the safety rope according to established guidelines. Follow these guidelines no matter what.
- Test natural anchors.
- Before and during the descent, if you are with someone, use established communication commands.
- Use two *carabiniers* (hooks) to hold sling and ropes.
- Inspect ropes and knots, leaving the rope uncoiled in a pile.
- When descending, position feet about 1 foot apart and make sure they do not get so high that they throw you off balance.
- Balance with your upper hand and brake with your lower hand.
- Lean forward from your hips with legs straight.
- As you rappel, inspect ridge for loose rocks.
- Don't bounce or stop suddenly, but rappel smoothly.
- When your descent is complete, remove rope, untie safety, and move quickly away from under the ridge in case of falling rocks.

Your family planned another camping trip for this weekend and allowed you to invite your best friend. The two of you have shared many adventures on picnics and campouts. You are anxious to show off your rappelling skills.

After breakfast, you are ready to begin hiking. Your parents advise against rappelling because your friend has never done this before. You or your friend could get hurt. Without your older brother or sister, you might get in a hurry and forget the rules. You are both disappointed but decide to go hiking anyway. You put rappelling gear in your backpack to get the feel of hiking with a full load.

As you wander, you are excited by the things in nature you see. You take your time and look over the beautiful landscape. You are aware that the grass and trees are very, very dry. Without watching the time, you suddenly notice it is late afternoon. You had promised to be back at the campsite by dark. You realize you are some distance from the campsite. In your enthusiasm, you forgot to mark your trail or to pay attention to landmarks to retrace your trail. In other words, you both forgot two important rules all hikers must follow.

You don't panic.

You say you thought you remembered the right direction but not the whole trail. After over an hour of walking, you come upon a familiar path. You have found a pathway back to the campsite but, unfortunately, it is in the opposite direction from where members of your family will be looking for you. Even worse, you are standing on the edge of a cliff 35 feet above the trail on a ridge. This cliff hangs directly over a small group of slanting, almost vertical rocks that separate you and your friend from the trail.

It will take you hours to find a way down to that trail. It is almost sunset. You know you are not prepared to camp overnight where you are now. You know you must act before it gets too dark.

Your friend begs you to demonstrate your rappelling skills, insisting that you rappel down the cliff and go for help. Your friend will wait alone, hoping you get down safely and find help soon. The rope you are carrying is long enough to reach the path below.

You could try to teach your friend how to rappel down the side of the cliff. You remember your parents' advice about not rappelling. There is no moon tonight and it will be pitch dark within a half an hour. You can feel the night chill already.

You quickly consider the situation. You accept that there are only five things you can do. You must consider which one is the best for you in this situation. You could

1. leave your friend on the cliff, rappel down, and go for help (however, if you hurt yourself, there will be no one to help you)
2. decide to stay with your friend, take turns yelling, and keep close together to stay warm
3. try to teach your friend the basic steps of rappelling, and then both of you try to rappel down the cliff
4. start a fire and hope that it attracts the attention of other hikers and campers through the dense but extremely dry grasslands, bushes, and trees
5. try to descend the cliff without the benefit of rappelling on the ropes, hoping to get close enough to the trail before dark to jump

Before you make your final decision, take time to carefully consider the information you have available. To help you review the important information, complete Predecision Task Sheets 1 and 2.

PREDECISION TASK SHEET 1

To help you review your situation, answer the questions below. (Try to answer first without looking at what you just read.) Return to the story only when you cannot remember specific details. Paraphrase the information rather than copying it down word for word.

1. Based on the story, what are seven important facts about proper rappelling?
   a.
   b.
   c.
   d.
   e.
   f.
   g.

2. Based on the story, what are seven important facts about proper hiking?
   a.
   b.
   c.
   d.
   e.
   f.
   g.

3. What are the four most important facts you know about the situation the two people in this story are in?
   a.

   b.

   c.

   d.

4. What are two ways the information concerning rappelling could be considered information about technology?
   a.

   b.

After you have completed your individual answers, share your response with the others in your group. Add to, revise, or correct your responses as needed. Make sure everyone in your group comprehends this information before going on to the next page.

PREDECISION TASK SHEET 2

Imagine you are one of the people in the story. Think about the situation. In the space below, write possible results of each action. Do this by yourself. Consult members of your group only after you have filled in all the spaces.

Choices	Possible Good Results of This Decision	Possible Bad Results of This Decision
1. Leave your friend and rappel yourself.	a. b.	a. b.
2. Stay with your friend.	a. b.	a. b.
3. Try to teach your friend how to rappel.	a. b.	a. b.
4. Start a fire and wait.	a. b.	a. b.
5. Descend the cliff without rappelling.	a. b.	a. b.

After you have completed your individual answers, share your responses with the others in your group. Add to, revise, or correct your responses as needed. Make sure everyone in your group comprehends this information before going on to the next page.

DECISION SHEET

1. After studying these choices, I believe my *best* decision is to

2. This is my best choice for these two reasons:
   a.

   b.

3. After discussing these with my friend, our *best* choice is to

   because

4. The *very worst* thing we could do in this situation is to

   because

After completing your decisions and writing them on this page, share your answers with the members of your group. In what ways are your answers identical? Similar? Different?

REVIEW AND REFLECTION QUESTIONS

Suggested follow-up questions to focus and guide thinking and learning.

a. From the context, what is a good definition of the term *rappelling?*

b. In the story, what reasons did you have to take along your rappelling gear?

c. In the story, why would your parents not want to you to rappel?

d. In the story, what safety rules did the friends not follow?

e. From the details in this story, would you say the two friends were actually hiking or just taking a nature walk?

f. From the details in this story, would it have been a good or bad idea to build a fire?

g. What does the word *safety* mean?

h. What are four things to remember about correct hiking?

i. What are four important things to remember about how to rappel?

j. How is rappelling a type of sport?

k. What makes hiking a sport?

l. In what ways did the friends in the story follow the safety rules?

m. If the two friends tried to rappel down the cliff, would they have been brave?

n. What are at least three good reasons for going hiking?

o. In terms of physical exercise, would hiking or rappelling be better exercise?

p. If the two stayed on top of the cliff all night, how could this act be called a safe one?

q. What does it mean to be safe?

r. If you were hiking or rappelling right now, what would be your feelings?

s. What are three things you could learn by going hiking? Rappelling?

t. In the story, the parents said no rappelling was allowed. If the friends had rappelled down the cliff and returned safely, what would have been the parents' reactions?

u. Suppose the boys had not rappelled and had stayed out all night long. What might the parents have said about rappelling?

v. If you found yourself in the situation in the story, what would have been the safest thing for you to do? The bravest thing?

w. What parts of this story involved science?

x. In what situations might scientists have to follow the guidelines for how to rappel?

y. To get the results you want in rappelling, you must consistently follow the guidelines and rules. In what ways is having to follow guidelines the same as what scientists must do in using a scientific method?

z. What types of scientists might need to rappel to conduct observations or complete their research?

Rank-Order Decision Episodes

Rank-order decision episodes provide students with a story in which a major character or group must make a decision from among listed options, each independent of the others. Students must consider the relative importance of a number of nearly identical alternatives and prioritize each. That students are made aware that their character's or group's choices will be considered by others in the order of their own rank is critically important. The procedure required to complete these episodes ensures that students will select the top option from the remaining alternatives all the way down to the final pair of options to be rank ordered.

EPISODE **8**

HEAVENLY BODIES

Astronomers and other people interested in space are upset. In recent years, Congress has drastically cut money for space investigations and exploration. Scientists in many different areas are upset because little new information about space can be discovered through Earth-based observations. Deep-space probes, fly-by missions, and landings are necessary to obtain large amounts of important new information. Astronomers have argued that there are still wonders of space that must be scientifically examined and explained. Information about the origin of the universe, Earth, and early life on Earth could be found with space flights.

Meanwhile, little publicity has been given to the cuts in the space program. Funding for space has been limited to a few projects, with the space shuttles receiving most of the money. Manned flights to the moon are not planned for this century. No new voyages to land a vehicle on Mars are being prepared. It seems the United

States is ready to drop its space efforts. Only the space shuttles, business communications satellites, and military weapons projects are being supported by Congress and the President.

You are an astronomer. Imagine you are trying to get support for new space flights. You have made speeches on radio and television to raise people's knowledge of space projects. You want Congress to fund many different space projects.

After months of work, you are able to get a few people to help you advertise the space programs. They are scientists and nonscientists who believe in the benefits of knowledge about space. They believe we can learn much about Earth and its past, present, and future with space missions. Working with them, you have made a list of projects you want Congress and the people to support. Below is a description of each project.

A. *Missions to explore sunspot activity.* Since Earth depends on heat and light from the sun, what happens on the sun is important to the climate and life on this planet. Sunspots are dark spots on the sun's surface. They are cool areas often surrounded by streamers of exploding and expanding gases reaching thousands of miles into space. These spots and streamers tell us about storms on the sun. They also serve as storm warnings for the solar system. When the number of sun spots are few, Earth has colder temperatures and ice glaciers move faster across Earth's surface.

B. *Missions to land instruments on Mercury.* Since Mercury is the closest planet to the sun, instruments could send a lot of information about the sun to Earth. Information about sun spots, the sun's gravitational pull, and solar winds could be valuable in predicting climate changes on Earth. Information about the soil, geography, and climate of Mercury could tell us about the history of the solar system.

C. *Missions to explore Venus.* Venus is the only planet that rotates backward. If you could see the sun through Venus' dark clouds, which are hundreds of miles high, the sun would rise in the west and set in the east. Venus's atmosphere is over 900° Fahrenheit. This exploration would include several landings on the surface and circling the planet with a system of satellites containing scientific instruments. Information about the soil, geography, atmosphere, and temperature could tell us a great deal. Some scientists believe Venus was not one of the original planets. Because of the carbon-based, dense clouds, Venus's surface has never been seen from Earth telescopes.

D. *Missions to explore Mars.* Mars seems to have conditions that may make it possible for humans to live there. Photographs, soil analysis, and laboratory tests conducted by the Viking team revealed no recognizable life forms. However, Mars is the most likely place for life as we know it to be found in our solar system. Scientists still want information about the Martian "canals," ice cap, soil composition, and dust storms. This mission would include a series of instrument landings and satellites. Humans would land on Mars by 2010.

E. *Missions to explore the asteroid belt.* This belt of asteroids is made up of small and large chunks of rocks. Some asteroids are 30 miles across. Sometimes called the *minor planets,* these asteroids lie between the orbits of Mars and Jupiter. They may be fragments of a large planet that exploded or collided with another space object. Information on these asteroids could help save Earth one day if a large asteroid approached our planet. Other information could tell us about the history of the solar system.

F. *Missions to land on Io.* Pictures taken from previous fly-by missions suggest the large moon of Jupiter, Io, is experiencing live, active volcanoes. Scientists believe the core of Io is very old, while the surface is young. Information from the surface could give details about Jupiter, its other moons, and other space objects such as meteorites and comets. Data collected on radiation, electricity, and volcanic activity might help us understand the early history of Earth. From Io, information about Jupiter, sometimes called a *dark star* or *near star,* could be obtained.

G. *Missions to fly by and orbit Saturn.* Information gathered from the rings of Saturn may help explain the origin of our solar system. Scientists question if these icy rings, which move at different speeds around Saturn, are "leftovers" from Saturn's creation. The rings may be part of a moon that exploded. Many other facts about the rings and the gaseous atmosphere of Saturn would be useful to understand the outer planets of the solar system.

H. *Missions to probe several large comets.* In 2062, Comet Halley will return on its 76-year orbit around the sun. It will pass near Earth. However, that is too long a period of time for scientists to wait to explore comets. Space probes could be sent close to the heads of several large comets within our solar system. These data would allow us to understand what makes up a comet and why comets behave as they do. Information about the comet's tail, which is sometimes millions of miles long, may help us understand many things about deep space. The tail may also contain chemicals, minerals, or life forms from outside our solar system.

Your group is proud of its list of important missions you want to have carried out. At present, these seem impossible as no money is available to start them. You are disappointed but not ready to quit.

This morning things are different. You received a phone call. A very powerful congresswoman heard your last speech. The congresswoman says that the space program should not die but should be kept alive with new projects and missions. She and her close friend, a congressman from a nearby state, will try to get money from Congress for space. She tells you we must explore the heavenly bodies to help us better live on Spaceship Earth.

She tells you that Congress could never support all eight projects at the same time. Instead, Congress might fund one mission, then a second, and so on. Each project would have to be successful before the next one could be funded. She says Congress could fund just one project or maybe even five or six missions, but never all eight. She wants you to help decide which missions Congress should fund. She also wants you to help decide the order in which these projects should take place. She will ask Congress to fund the project you help pick as the most important one. You are told that if the first mission goes successfully, Congress will grant money for your second mission. If this mission goes well, your third mission will be funded. In other words, money will be made available for the projects in the order you rank them, until Congress has no more money to spend for space missions.

Which heavenly body will you select as the one to explore first? Second? Third? You are to make your decision as a member of a small group of citizens.

Before you make your final group decision, complete Predecision Task Sheets 1 and 2. These will help you think about each project in important ways. Make sure all members of your group complete these Task Sheets by themselves before sharing answers with the group. Then discuss what you found out about each heavenly body with the members of your group.

Once this is done, complete the Individual Decision Sheet. This sheet is where you record your personal decisions. Make your own decisions first. Then, and only then, meet with your group to make a final group decision. Record your group's final decisions on the Group Decision Sheet.

PREDECISION TASK SHEET 1

Working on your own, fill in the chart below. Find information in other books or references to help with your answers.

Heavenly Body/Bodies	Two Facts We Know About This Object in the Solar System	Two Ways Information About This Object Could Help People on Earth
1. Sun	**a.**	**a.**
	b.	**b.**
2. Mercury	**a.**	**a.**
	b.	**b.**
3. Venus	**a.**	**a.**
	b.	**b.**
4. Mars	**a.**	**a.**
	b.	**b.**
5. Asteroid belt	**a.**	**a.**
	b.	**b.**
6. Io, Moon of Jupiter	**a.**	**a.**
	b.	**b.**
7. Saturn	**a.**	**a.**
	b.	**b.**
8. Comets	**a.**	**a.**
	b.	**b.**

After you have completed your individual answers for the chart, share your responses with your group. Add to, revise, or correct your responses as needed. Make sure everyone in your group comprehends this information before going on to the next page.

PREDECISION TASK SHEET 2

Information we can gather about space and objects in space can help people on Earth in different ways. Scientists help to plan space missions and probes, to build space rockets, and to study the information from space missions. Answer the following questions by yourself before meeting with members of your group.

1. An astronomer is a scientist. What are two things an astronomer might do that tells you he or she is a scientist?
   a.

   b.

2. In the story, a scientist is trying to get people to support one area of science. Should scientists try to get money from Congress for their projects?

   What are the two strongest reasons for your decision?
   a.

   b.

3. Suppose someone said scientists should worry about and study what is on Earth and not study heavenly bodies. What would you say to this person?

4. Suppose a group of scientists opposed missions into space. Would these persons still be considered scientists? _________

   What are two important reasons some scientists would favor new missions to space while others would not?
   a.

   b.

5. At what point does a space mission become a science mission?

6. In addition to astronomy, what are at least five areas of *science* that would be interested in space objects?
   a.

   b.

   c.

   d.

   e.

7. What are at least five areas of *technology* that would be used to complete these eight missions?
   a.

   b.

   c.

   d.

   e.

After you have completed your individual answers for this sheet, share your responses with the others in your group. Add to, revise, or correct your responses as needed. Make sure everyone in your group comprehends this information before going on to the next page.

INDIVIDUAL DECISION SHEET

Imagine that Congress has finally agreed to fund at least one space mission. If successful, it may spend money for a second, third, or fourth mission. It will not be spending money for all of the missions. Working on your own, rank order these eight missions according to their importance to the future of humankind. Place a 1 in the blank for your first choice, a 2 for your second choice, and so on. An 8 will mark your least important mission. Keep in mind that if your first project goes well, Congress will give money for the second-ranked project. This order of funding will continue until Congress has no more money to spend for space missions.

Rank	Mission Description
_____	Mission to Explore Sunspot Activity
_____	Mission to Land Instruments on Mercury
_____	Mission to Explore Venus
_____	Mission to Explore Mars
_____	Mission to Explore the Asteroid Belt
_____	Mission to Land on Io, Jupiter's Moon
_____	Mission to Fly By and Orbit Saturn
_____	Mission to Probe Several Large Comets

1. The reasons I selected the mission to

____

as the *most important* space project for humankind are

a.

b.

2. This project would be most helpful if it gave me information about

a.

b.

c.

3. This mission will help people on Earth in these three important ways:
   a.

   b.

   c.

4. Congress should fund space explorations such as these because
   a.

   b.

   c.

5. If Congress would fund any space project I wanted, I would want them to spend money for a mission to

   ____

   because

GROUP DECISION SHEET

Imagine that Congress has finally agreed to fund at least one space mission. If successful, it may spend money for a second, third, or fourth mission. It will not be spending money for all of the missions. Discuss the space missions with members of your group. Arrive at a decision using consensus rather than by voting with majority rule. *As a group,* rank order these eight missions according to their importance to the future of humankind. Place a 1 in the blank for your first choice, a 2 for your second choice, and so on. An 8 will mark your least important mission.

Rank	Mission Description
_____	Mission to Explore Sunspot Activity
_____	Mission to Land Instruments on Mercury
_____	Mission to Explore Venus
_____	Mission to Explore Mars
_____	Mission to Explore the Asteroid Belt
_____	Mission to Land on Io, Jupiter's Moon
_____	Mission to Fly By and Orbit Saturn
_____	Mission to Probe Several Large Comets

1. The reasons we selected the mission to

____

as the *most important* space project for humankind are

a.

b.

2. This project would be most helpful if it gave us information about

a.

b.

c.

3. This mission will help people on Earth in these three important ways:
   a.

   b.

   c.

4. Congress should fund space explorations such as these because
   a.

   b.

   c.

5. If Congress would fund any space project we wanted, we would want them to spend money for a mission to

   ____

   because

REVIEW AND REFLECTION QUESTIONS

Suggested follow-up questions to focus and guide inquiry and learning.

a. What does an astronomer do?

b. What makes an astronomer a scientist?

c. What kinds of equipment might astronomers use to study space?

d. Why would scientists use these kinds of equipment?

e. What things might astronomers find on Earth that would be useful in their study of space?

f. What is a *heavenly body?*

g. What explanations did you find as to why Venus rotates in a different direction from all other planets in our solar system?

h. What is a *solar system?*

i. Scientists have recently begun to study more data on the atmosphere and surface of Venus. What information did you find on these recent discoveries?

j. Jupiter has been referred to as a *dark star.* What is a dark star?

k. Of the objects in space in the story, which one would you most like to visit?

l. Of the objects in space in the story, which one would you never want to visit?

m. Why would scientists be interested in objects in space?

n. What is an *astrologer?*

o. What are two ways an astronomer is different from an astrologer?

p. What are two ways an astronomer is like an astrologer?

q. What is a *horoscope?*

r. Scientists in many fields have found that horoscopes are not scientific and are highly inaccurate. Yet millions of people still believe in their horoscopes. Why would people continue to believe in such things?

s. What is *space?*

t. In what ways is an astronaut a scientist? Not a scientist?

u. Why would some humans want to explore space and other humans not want us to explore space?

v. Space has been called the "final frontier." What is a *frontier?*

w. In what ways is space a frontier?

x. In what ways are scientists and explorers identical? Similar? Different?

y. What scares you most about our exploration of space?

OOPS! THE ONE TIME I FORGOT!

You have developed a keen interest in microscopic organisms. Until recently, all your observations have revolved around prepared slides your parents bought you. You decided to share some of the slides on cells and tissues with your science class. Ms. Brackin, your teacher, understands your enthusiasm for this area of science. She arranged for you to meet the biology teacher, Ms. Szesze. Ms. Szesze invited you to the biology lab after school. She wanted to help you learn how to make your own slides and improve your charting methods. You agreed to work one afternoon a week with her.

You share all your information, interest, and enthusiasm with your parents. They have agreed to help you buy a better microscope, additional slides and covers, and a variety of containers and chemicals. You were delighted. You have budgeted your time. Each night you have at least one hour for your own microscopic work.

OOPS! THE ONE TIME I FORGOT!

As you work, you constantly keep in mind what Ms. Szesze said about keeping accurate records.

Records or notes are ways you know what you've done and how you've worked to get your results. Without such notes, you may not remember important details, facts, or steps in your experiment. You can also use notes to discover if you did something wrong. For these reasons, you must make sure your written records of your procedures and materials are accurate and complete.

When taking notes about an experiment or observation, you should write the major purpose for what you are doing or trying to do. You should accurately describe what you see or use (for example, names of chemicals, amount of materials used, sizes of containers). Then you should record the steps you follow during the experiment. Ms. Szesze stressed the importance of such listings for you to accurately repeat the experiment, should you want to do so. You know taking notes is not fun and takes time, but they are as important as the experiment itself. Finally, when the experiment is complete, a good scientist reviews the notes and records. These are relevant to the completed events and findings.

You are excited about science. Your care in following the right procedures in conducting experiments pleased your parents and teachers. Your charting records received praise from Ms. Szesze. Even more exciting was the news of the science fair, where you could show others what you have done. Besides the award, your parents promised you a much larger microscope should you win. For the past two months you have been working very hard on your science fair project.

Last night your parents insisted you get to bed earlier than usual. To meet their request, you failed to clean all of your equipment. In addition, you did not take the time to accurately write down your present experiment. Critical steps you took during the experiment were not described. You also failed to describe accurately the exact amounts of certain chemicals you used.

This morning you noticed organisms moving in the container beside your microscope. They appeared to be multiplying and growing larger. Under the microscope, you observed hundreds of single-celled organism with a small finlike structure on the side, a hexagon-shaped structure at the top, and a hornlike protrusion at the base. You have never seen or heard of this type organism before. You may have discovered a new organism. Surely this "discovery" will win you the award at tomorrow's science fair. The micro-organisms in this overnight project would be much better to display than the project you were ready to turn in.

All day you have thought about what you could do about the science fair and your discovery. You consider that you really have only four possible choices to make.

1. You could ignore the new discovery and turn in your original report and project. You probably won't win with this project, but you enjoyed doing it.
2. You could turn in the new discovery project and report how you accidentally found the new organisms. There would be only incomplete records and no charting records of the night before your discovery. The judge would frown upon the incomplete records. You will surely lose the science fair.
3. You could turn in your new discovery project with the theme "Why a Scientist Should Keep Accurate Records." This will not win the award, but it will remind others to follow charting methods in the future.
4. You could make up a set of charts that would be acceptable even though they would be incorrect and wrong. You know enough about charting that you could easily make up a chart record that would make sense to most judges. If you did this, your chance of winning is great.

Before you make your final choice, you decide to think about which choices are better than other. The Predecision Task Sheets that follow should help you with your final decision. Then use the Decision Sheet to record your final decision.

PREDECISION TASK SHEET 1

Before you consider the choices you might make, take time to review some things about scientists and their work. Review the story for answers to the following questions. Complete this task by yourself.

1. What are three important reasons scientists make careful and accurate notes and records of what they observe, do, and use?
   a.

   b.

   c.

2. What are three things scientists write about in their notes?
   a.

   b.

   c.

3. What are ten adjectives you would use to describe the notes and records scientists should keep?

4. How would you define the term *science?*

5. In your own words, what does a scientist do?

6. Suppose a scientist was found to have made inaccurate records or to have made up procedures. How should other scientists view this scientist and his or her discoveries?

7. To what extent is it good that scientists take careful notes so they can repeat their experiments?

8. Suppose a scientist reported he or she used certain steps and ingredients in an experiment. Suppose this scientist received a great deal of praise for the results he or she achieved in the experiment. Suppose further that when you used these exact steps and ingredients, you failed to get the same results as the scientist reported. If this were to happen, what would be your evaluation of the results the scientist reported?

9. In your own words, what does it mean to do research?

10. What does it mean to do an experiment?

After you have completed your individual answers, share your responses with the others in your group. Add to, revise, or correct your responses as needed. Make sure everyone in your group comprehends this information before going on to the next page.

PREDECISION TASK SHEET 2

Complete this task on your own. Below are the four possible choices you have at the present time. Consider each one carefully. Write down the two best and worst reasons you might choose each of these possible decisions.

Decision	The Best Reasons to Choose This Option	The Worst Reasons to Choose This Option
1. Ignore the new discovery and turn in your original report and project.	a. b.	a. b.
2. Turn in the new discovery and report how you accidentally found the new organisms.	a. b.	a. b.
3. Turn in the new discovery with the theme, "Why a a Scientist Should Keep Accurate Records."	a. b.	a. b.
4. Make up a set of charts that would be acceptable even though they were incorrect and incomplete.	a. b.	a. b.

After you have completed your individual answers for this sheet, share your responses with the others in your group. Add to, revise, or correct your responses as needed. Make sure everyone in your group comprehends this information before going on to the next page.

INDIVIDUAL DECISION SHEET

Here are four decisions you could make. These seem like the only choices you have. Decide which is your best choice and place a 1 on the space next to it. Place a 2 next to your second-best choice, and so on.

______ Ignore the new discovery and turn in your original report and project. You may not win with this project, but you enjoyed doing it and you did it correctly.

______ Turn in the new discovery project and report how you accidentally found the organisms. There would be only incomplete records and no charting records of the night before your discovery. The judge would frown upon the incomplete records. You will surely lose the science fair.

______ Turn in your new discovery project with the theme "Why a Scientist Should Keep Accurate Records." This will not win the award, but it will remind others to follow charting methods in the future.

______ Make up a set of charts that would be acceptable even though they would be incorrect and wrong. You know enough about charting that you could easily make up a chart record that would make sense to most judges. If you did this, your chance of winning is good.

1. Two important reasons I picked choice___ as my *best* choice are

a.

b.

2. Two important reasons I picked choice ___ as my *worst* choice are

a.

b.

3. If I could do anything I wanted in this situation, I would

4. I would do this for the following three reasons:

a.

b.

c.

GROUP DECISION SHEET

Imagine that the members of your group are working together on a group project. Imagine further that the same thing happened in your group project that happened in this story. Therefore your entire group needs to reach an agreement on the best choices to make in this situation. Following are four possible decisions your group could make. At this time these seem like the only choices you have. Decide which is your best choice and place a 1 on the space next to it. Place a 2 next to your second-best choice, and so on.

______ Your group should ignore the new discovery and turn in your original report and project. You may not win with this project, but you enjoyed doing it and you did it correctly.

______ You should turn in the new discovery project and report how you accidentally found the organisms. There would be only incomplete records and no charting records of the night before your discovery. The judge would frown upon the incomplete records you would turn in. You will surely lose the science fair.

______ You could turn in your new discovery project with the theme "Why a Scientist Should Keep Accurate Records." This will not win the award, but it will remind others to follow charting methods in the future.

______ You could make up a set of charts that would be acceptable even though they would be incorrect and wrong. You know enough about charting that you could easily make up a chart record that would make sense to most judges. If you did this, your chance of winning is good.

1. Four important reasons we picked choice ______ as our *best* choice are
   a.

   b.

   c.

   d.

2. Four important reasons we picked choice _____ as our *worst* choice are

a.

b.

c.

d.

3. If we could do anything we wanted in this situation, we would

4. We would do this for the following three reasons:

a.

b.

c.

REVIEW AND REFLECTION QUESTIONS

Suggested follow-up questions to focus and guide inquiry and learning.

a. According to this story, why is it important to keep good records of one's observations?

b. What is a good definition of a *discovery?*

c. What does it mean to "chart one's observations"?

d. What does a scientist do?

e. What are three recent discoveries scientists have made?

f. What is a *microorganism?*

g. What is one major point this story is trying to make?

h. If you were the person in the story, would you call yourself a scientist?

i. What are three areas in science you would like to explore?

j. If you could make one new discovery in science, what would it be?

k. Suppose you made a big discovery but forgot to keep accurate records. What word would best describe your feelings?

l. What is one difference between an experiment and an observation?

m. In what ways might a science project in school be similar to a project conducted in a scientist's laboratory?

n. In the story, what was the project for the person to turn in for the science fair?

o. Suppose a person turned in made-up records and someone found out about it. What would people think about other projects the person did?

p. If you were the person in the story, what steps would you take to *rediscover* these microorganisms?

q. Is it good or bad that scientists are so careful in recording their activities?

r. What would it take for you to become a scientist?

s. What would the world be like without the discoveries of science?

t. In the story, the person was concerned about winning the science fair. Suppose the project with the made-up records won. In what ways would this project be labeled a *science* project?

u. If such a made-up project won the science fair, in what ways would this award have been fair?

v. If a scientist is found to have made up his or her data for an experiment, what should other scientists do to this scientist? To the scientist's data?

w. What would it take for you to make accurate records of the procedures you used during every science project?

BECOMING A MASTER SCIENTIST

Assume you are an aspiring young scientist working and studying under a master scientist at a major research laboratory in Europe. You applied for this position because you believed a good scientist is one who has had intense exposure to laboratory work, to libraries, and to many talented scientists. You believe a scientist should be skilled in many research techniques. You believe imagination and curiosity are essential to doing good science. You feel scientists should freely explore their innermost ideas and hunches within the areas they prefer to investigate. You also believe scientists must be more concerned with *how* to investigate things rather than *what* objects or subjects they are studying.

When you applied for the position of resident student-scientist in the "Study Abroad with a Master Scientist" contest, you felt you already possessed many qualities of a great scientist.

BECOMING A MASTER SCIENTIST

You were informed that the Future Scientist Search Committee selected you because of your philosophy about scientists rather than your specific knowledge and abilities. The committee stated they wanted to assist you in developing techniques that would help make you an accomplished scientist. In other words, they wanted you to have the opportunity to become skilled in the abilities to go along with your beliefs about science.

Upon arriving in England, the master scientist welcomed you into her own home. Over the past several months you have observed her in her private office and lab. She has answered all your questions. She seemed to be all the things you thought a master scientist should be. She is dedicated and well disciplined. She is obviously skilled and possesses great imagination and knowledge in many fields. Her works demonstrate her ability to translate her inner curiosity into high-quality scientific experiments, data, and conclusions. She has suggested numerous ways you can become just as skilled as she.

This morning you began your first formal research project under her direction. To your surprise, she was very critical of your effort. She said that what you call "scientific inquiry and objectivity" is what she calls "trial-and-error guesswork lacking acceptable scientific methods." She said, "True scientific inquiry is possible only when one's work shows skillful and purposeful use of the techniques and concepts scientists use—techniques and concepts you do not now possess. Scientific work involves few clever or quick conclusions. Great scientists never ignore scientific procedures and logical reasoning from the beginning to the end of an experiment—as you have just done. Your talk of scientific accomplishment must wait until you have mastered basic inquiry and research techniques. You must learn patience, precision, attention to detail, and to follow procedures."

Furthermore, she insisted you conduct future investigations as she directs you to. Until further notice, you will do science as she tells you to do it. Nothing less will be tolerated!

You admire this master scientist for her talents and accomplishments. However, you are upset by her lack of tolerance for your curiosity and innovations in inquiry and efforts to "do" science. You know exposure to such a great scientist will be of tremendous future value to you. In fact, to stay with her would enhance your chances of getting a scholarship to a major university and of finding a high-paying scientific job when you return home after a year. Her intolerance angers you! You believe to continue to stay with her would violate the philosophy that won you this opportunity to study in England.

When you left the laboratory this morning, you were upset, angry, and confused. Now you know you need to examine your thoughts and feelings. You need to reflect upon your view of science and the view the master scientist has of science. To help you in this effort, answer the questions below.

1. From the paragraphs above, what is your view of science and of doing scientific investigation?

2. What is the view of science and of doing scientific investigation that the master scientist holds?

3. In what ways are these views identical?

4. In what ways are these views different?

5. What are at least three reasons for these differences?
   a.

   b.

   c.

Before going on, take time with members of your group to review the situation so far. Share your responses with others in your group, adding other answers to those you have written. Do not go any further until everyone in your group can accurately describe the situation and give adequate answers to the questions above.

You make a list of the options you believe you have open to you. You decide you will study the list and decide what to do from among the alternatives you have identified. Your list contains five possible courses of action. The five choices are

1. I could do as she says because I came to England to study how a scientist does research. This is a once-in-a-lifetime experience I cannot afford to miss. However, to stay would mean I have to accept and follow all the rules and regulations set down by this master scientist. I cannot do science as I want. Rather I will learn to become skilled in the ways this one scientist conducts scientific research.
2. I could refuse to allow this one scientist to take my freedom of expression away from me. I would demand that I be left alone to do my own thing. If she refuses, I will immediately leave the laboratory and program and return home.
3. I could attempt to explain again to the master scientist what my philosophy is and what I believe makes a good scientist. However, she has already refused to hear my "nonsense." To bring this matter up again would only anger her more. I know she will be infuriated. My own beliefs are what count.
4. I could do what the master scientist tells me to do during the supervised study and laboratory time. I would spend my evenings investigating what and how I wish. This would at least allow me to develop some new abilities while letting me do my own thing. However, this would violate the rules of the program. If I am caught, I will be sent home.
5. I could write a letter to the Future Scientist Search Committee and explain my side of the situation to them and demand (or request) that I be placed with another scientist. However, I have heard that in the past when such things were done by other students, the committee settled the problem by sending the student home.

Before you rank order these possible courses of action, you decide to review the situation. You investigate the situation as a scientist might study a problem. To help you in this inquiry, answer the questions on the Predecision Task Sheet that follows. Only after you have finished your inquiry will you make your final decision. Use the Decision Sheet to record your final answer.

PREDECISION TASK SHEET

Write answers to the following questions that best reflect your thoughts and feelings in this situation.

1. In this situation, what is the problem?

2. What are at least two important reasons this is the problem?
   a.

   b.

3. What are the causes of this problem?

4. In this situation, what do you most want to gain from this experience of working with a master scientist?

5. In this situation, what are you most willing to give up to resolve the problem you have?

6. What are at least three possible solutions to this problem?
   a.

   b.

   c.

7. In this situation, what decisions would expert scientists expect you to make?

8. What alternative is most likely to work?

9. In this situation, how would your preferred solution most likely affect the scientist you are working with?

10. What are the two best reasons this alternative is the most likely to work?
    a.

    b.

11. To make this option work to resolve the problem, what specific actions must you take?

Once everyone in your group has answered these questions on his or her own, share your answers. Clarify and give reasons for the answers as your group considers each question. Add other answers to those you already have. Do not go on to make your final decision until your group has carefully studied and discussed the answers to all these questions.

INDIVIDUAL DECISION SHEET

Working on your own, consider the five options below *as the only ones open to you in the situation.* You decide you will try one option at a time, moving to the next if one doesn't work. First you need to decide which of the five options is the *best* to follow down to the *worst.* Consequently you decide to rank the five options. You will place a 1 by the best choice, a 2 by the second-best choice, and so on until you have ranked all the options. Remember, the option you rank as second will be followed if and only if the first one is not successful. Your list contains the following alternatives:

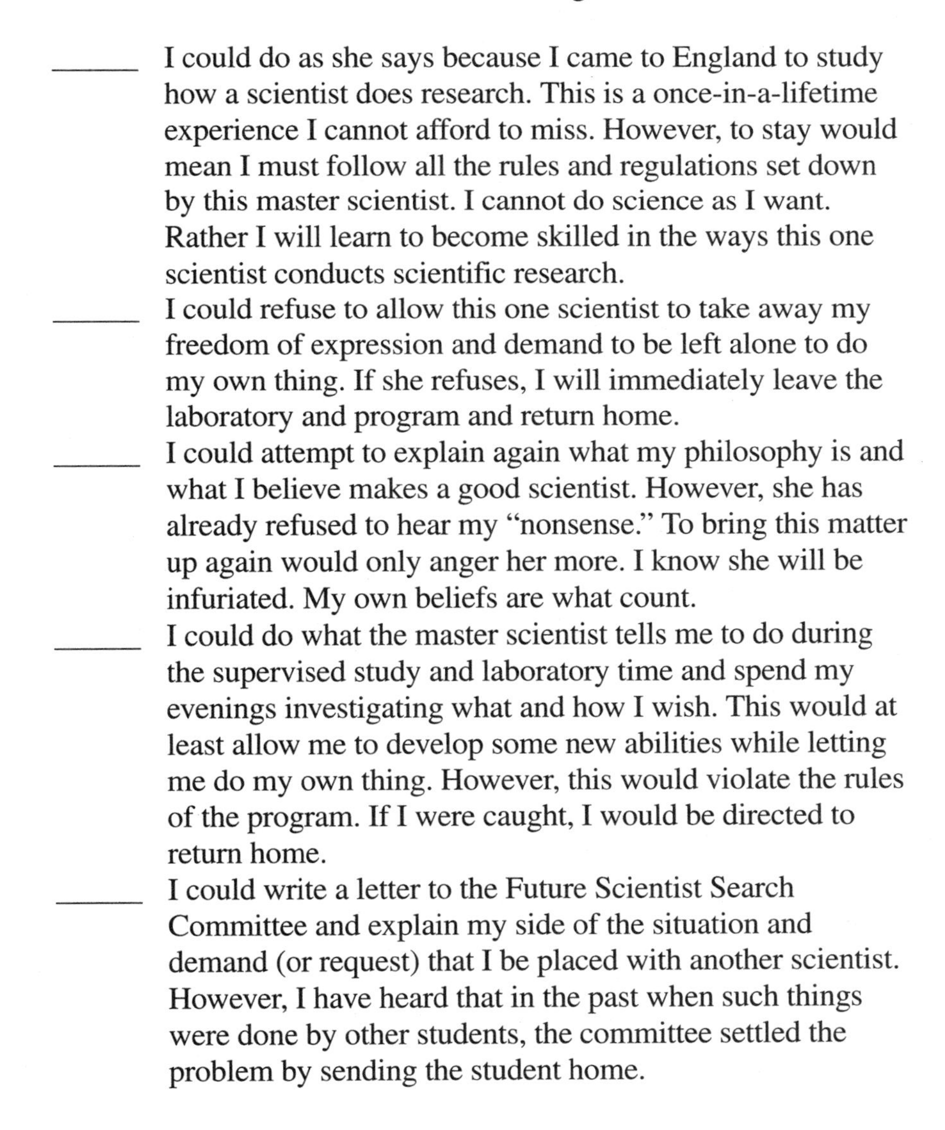

______ I could do as she says because I came to England to study how a scientist does research. This is a once-in-a-lifetime experience I cannot afford to miss. However, to stay would mean I must follow all the rules and regulations set down by this master scientist. I cannot do science as I want. Rather I will learn to become skilled in the ways this one scientist conducts scientific research.

______ I could refuse to allow this one scientist to take away my freedom of expression and demand to be left alone to do my own thing. If she refuses, I will immediately leave the laboratory and program and return home.

______ I could attempt to explain again what my philosophy is and what I believe makes a good scientist. However, she has already refused to hear my "nonsense." To bring this matter up again would only anger her more. I know she will be infuriated. My own beliefs are what count.

______ I could do what the master scientist tells me to do during the supervised study and laboratory time and spend my evenings investigating what and how I wish. This would at least allow me to develop some new abilities while letting me do my own thing. However, this would violate the rules of the program. If I were caught, I would be directed to return home.

______ I could write a letter to the Future Scientist Search Committee and explain my side of the situation and demand (or request) that I be placed with another scientist. However, I have heard that in the past when such things were done by other students, the committee settled the problem by sending the student home.

1. If I were asked to explain or justify my first choice, I would say

2. The three things I would *gain* most from this first course of action are
   a.
   b.
   c.

3. The three things I most want to *avoid* by my actions are
   a.
   b.
   c.

4. This decision will affect my abilities as a scientist in the following three ways:
   a.
   b.
   c.

5. This decision will affect my future as a scientist in the following three ways:
   a.
   b.
   c.

6. My fifth-ranked option was the *worst* decision I could make for the following three reasons:
   a.
   b.
   c.

GROUP DECISION SHEET

As a group, consider the five options listed below *as the only ones open to you in the situation.* Try one option at a time, moving to the next if one doesn't work. First you need to decide which of the five options is the *best* to follow down to the *worst.* Consequently you decide to rank the five options. You will place a 1 by the best choice, a 2 by the second-best choice, and so on until you have ranked all the options. Remember, the option you rank as second will be followed if and only if the first one is not successful. You are to reach this decision by consensus and everyone must agree on all the rankings and the reasons given for these choices.

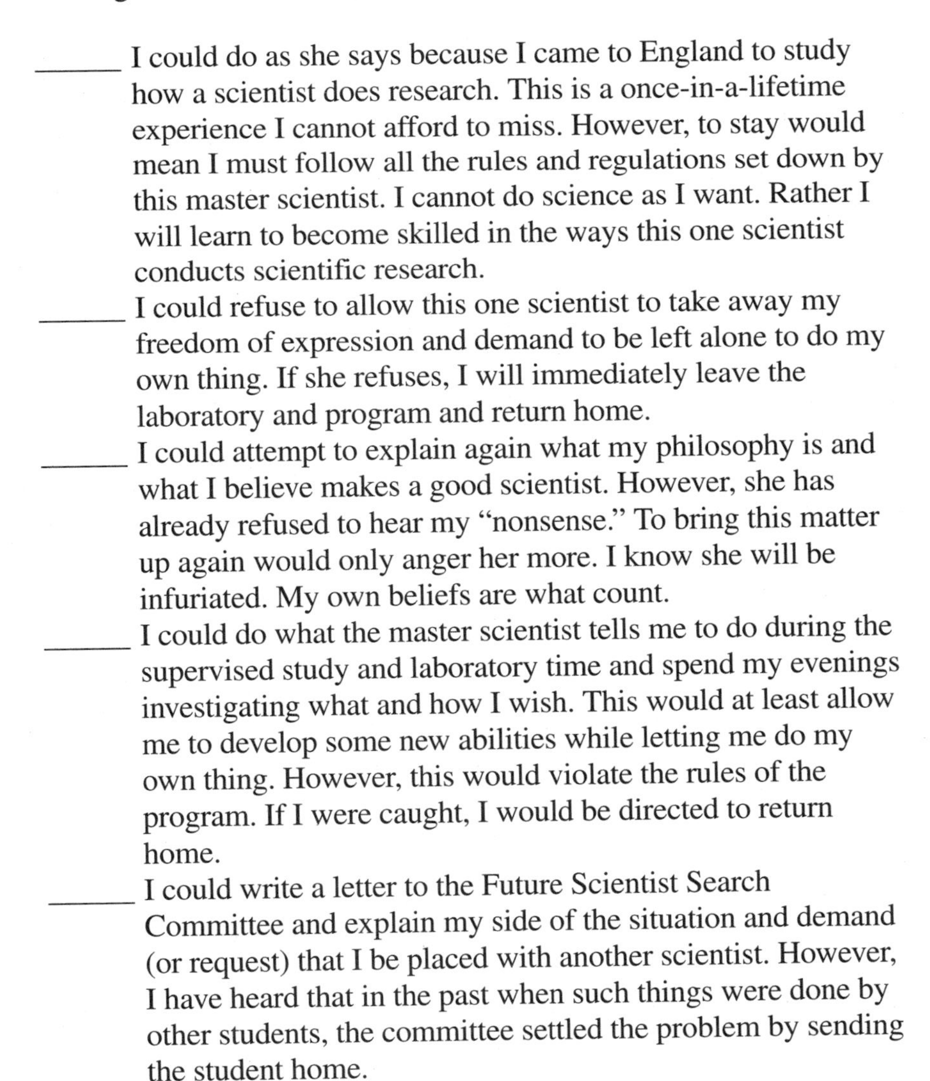

______ I could do as she says because I came to England to study how a scientist does research. This is a once-in-a-lifetime experience I cannot afford to miss. However, to stay would mean I must follow all the rules and regulations set down by this master scientist. I cannot do science as I want. Rather I will learn to become skilled in the ways this one scientist conducts scientific research.

______ I could refuse to allow this one scientist to take away my freedom of expression and demand to be left alone to do my own thing. If she refuses, I will immediately leave the laboratory and program and return home.

______ I could attempt to explain again what my philosophy is and what I believe makes a good scientist. However, she has already refused to hear my "nonsense." To bring this matter up again would only anger her more. I know she will be infuriated. My own beliefs are what count.

______ I could do what the master scientist tells me to do during the supervised study and laboratory time and spend my evenings investigating what and how I wish. This would at least allow me to develop some new abilities while letting me do my own thing. However, this would violate the rules of the program. If I were caught, I would be directed to return home.

______ I could write a letter to the Future Scientist Search Committee and explain my side of the situation and demand (or request) that I be placed with another scientist. However, I have heard that in the past when such things were done by other students, the committee settled the problem by sending the student home.

1. If we were asked to explain or justify our first choice, we would say

2. The three things we would *gain* most from this first course of action are
   a.

   b.

   c.

3. The three things we most want to *avoid* by our actions are
   a.

   b.

   c.

4. This decision will affect our abilities as scientists in the following three ways:
   a.

   b.

   c.

5. This decision will affect our future as scientists in the following three ways:
   a.

   b.

   c.

6. Our fifth-ranked option was the *worst* decision we could make for the following three reasons:
   a.

   b.

   c.

REVIEW AND REFLECTION QUESTIONS

Suggested follow-up questions to focus and guide inquiry and learning.

a. What do scientists do when they are *doing* science?

b. How might a scientist define *scientific inquiry?*

c. What would a scientist have to do to become known as a great scientist?

d. Do scientists become recognized for what they discover or for the methods they use in the discovery process?

e. What are at least three things you believe about science?

f. What are at least three things you believe about scientists?

g. What are at least three things you believe about how scientists complete their investigations?

h. When someone does *research,* what does that person actually do?

i. What is at least one major difference between research and investigation?

j. In what ways are research and investigation identical?

k. To conduct a successful scientific investigation, how young can a person be?

l. At this moment, how prepared are you to begin a scientific investigation?

m. What makes you believe you are as prepared to do scientific research as you say you are?

n. To become better prepared to complete a successful scientific investigation, what would you need to do? What would keep you from starting such a research project?

o. What would keep you from successfully completing a research project you might start in the near future?

p. As scientists are doing their research, how might they feel about the work they are doing?

q. To what extent is it good that scientists are excited about the work they do?

r. Many scientists spend a great deal of time studying books and other reading materials as part of their research. Does the fact they read so much turn you on or off about being a scientist?

s. How might a person's feelings affect his or her ability to complete a science investigation?

t. How might a person's feelings affect his or her ability to collect objective information about what he or she is studying?

u. If scientists realize their feelings are affecting the way they do research, what should they do about those feelings?

v. What does this story reveal about the ways doing science is affected by one's society or a social group?

w. In what situation is a person free to ignore a society in order to do science?

x. Suppose a person said that a scientist is never free of his or her society and its influences. What are the reasons this statement is true? False?

EPISODE **11**

CIRCLE OF POISON

You are a member of the President's Special Committee on the Use of Pesticides. Your committee has been holding meetings to discover why people are getting sick and dying from pesticides. The committee is expected to propose a policy that would help stop such illness and deaths in the future. To date, the committee has heard a number of witnesses. One topic that was investigated concerned exports of deadly pesticides. Important officials from inside and outside the government made the following statements:

Official 1

"Every hour millions of Americans eat fresh fruit, vegetables, and meat that contain cancer-causing pesticides. These foods are not grown in the United States. The U.S. has one of the safest food supplies and home-grown food-inspection systems in the world. These dangerous foods are imported from other nations. Nearly

30 percent of the food Americans eat is imported from nations that use pesticides that are illegal to use in the United States."

Official 2
"Federal inspectors have found high levels of pesticides in beef from Honduras, pineapple products from Thailand and the Philippines, and beans and carrots from Latin American nations. Eight percent of imported pears contained dangerous American-made pesticides. Imported peas, eggplants, cabbage, and peppers were among other vegetables with these chemicals on or in them. When these chemicals are in the foods, they cannot be washed away."

Official 3
"According to congressional testimony, the U.S. Food and Drug Administration allows virtually all imported foods to enter into this country without inspection. A major reason for this lax policy is to keep food prices low for American consumers. Americans want cheap food and so we accept what other nations send us."

Official 4
"Yes, we want food that costs less but we don't expect it to poison us! Cancer-causing pesticides show up on food at the grocery store because of a loophole in our laws. The current laws allow scientifically produced pesticides made in the U.S. to be sold overseas to any nation who can buy them. However, it is illegal for Americans to use these same poisons for crops grown here.

"Since 1983, one large U.S. chemical company has sought to have the Environmental Protection Agency approve the sale of Gallant *(haloxyfop-methyl)* for use in the U.S. The EPA has refused because of strong evidence that connects this pesticide with cancer. However, Thailand, our major foreign source of pineapples and pineapple juice, still buys tons of Gallant from this company.

"Between 1987 and 1989, Velsicol exported nearly 50 million pounds of Chlordane and Heptachlor to 25 different nations. Yet, before 1980, the EPA stopped all use of these two chemicals in the U.S. because of scientific research that linked both to cancer.

"Americans companies make billions of dollars each year selling known cancer-causing pesticides to other nations. American workers justify this by saying they will lose their jobs if they can't continue to export these pesticides to other countries."

Official 5
"What we have here is a 'circle of poison.' We export poisons to other nations that use them to spray their crops. Then we buy these

crops and eat them nearly every day of our lives. What good is it to ban these pesticides from use in this country and then buy foods sprayed with these same poisons from other countries? We not only poison the foods these people eat, we poison ourselves."

Official 6

"The problem is actually worse than the figures indicate. Since the Food and Drug Administration (FDA) does not routinely check and test food for pesticide contamination, the percentage of imported food that contains these poisons is likely to be higher than the 8-percent level reported by the government. Only 1 percent to 2 percent of imported food is actually tested by the FDA. Yet, by law, it is supposed to check almost all imported foods.

"The problem is getting worse. During the 1980s, the demand for pesticides doubled worldwide. The American exports amount to over a quarter of all pesticides sold in the world market. In addition, these are often sold to nations whose people are not always experts in their correct use. The United Nations claims that 115 countries do not have the technical expertise to use pesticides correctly. Add to this problem the fact that many pesticides they do use are very dangerous to humans. Anyone can see that they are poisoning themselves at an increasing rate."

Official 7

"If these pesticides are unsafe for American consumers, they should be banned from overseas sales. It does no good to ban these chemicals in this country and then buy foods produced with these chemicals from overseas. The EPA should rule to stop the manufacture and export of these cancer-causing pesticides immediately. Only by doing this will the 'circle of poison' end."

Official 8

"The problem is not in our sale of these pesticides to other nations. Most of these nations are poor and cannot afford to make safe pesticides for themselves. What we must do is increase the efforts of the FDA to do a better job of inspecting the food we import. We must educate Americans to wash the foods they eat even better than they currently wash them. It is unfair to blame the pesticide companies for this 'circle of poison.' To stop the sale of these pesticides overseas is to stop science and put people in this nation out of work. No government agency should be allowed to do this."

Your committee has heard the testimony. Now it is time to consider the data and make a recommendation. There is a good chance your recommendation will be accepted as official government policy. In this situation, you are asked to select only one of five choices as the policy your committee approves. These choices are

1. Allow continued exports of these pesticides and urge Americans to wash their foods more thoroughly.
2. Allow continued exports of these pesticides and increase inspection of all foods as they enter this nation.
3. Allow continued exports of these pesticides and stop all imports of foods from nations that use these chemicals.
4. Stop all manufacture and exports of these pesticides after 10 years to allow manufacturers to discover new pesticides.
5. Stop all manufacture and exports of these pesticides immediately.

Before you make your final decision as a group, you decide each of you will make your own decisions first. Then you will move toward a consensus decision. The steps you will follow are

- As a group, review the important information provided by the officials. Make sure all members are knowledgeable about what these officials have told the committee.
- Complete the Predecision Task Sheets.
- Review with the group the responses you have made to the items on the Predecision Task Sheets.
- Make your final personal decision and record it on the Individual Decision Sheet. Each person is to rank order all the possible options. Whatever option is ranked 1 will be tried first. If that fails, whatever option is ranked 2 will be tried. This will continue until the problem stops. Remember, even when you are down to the last two options, select the one you believe will work the best of those two. Therefore, it is important how each person ranks each option.
- Finally, work with members of the committee as a group to arrive at a consensus decision. The group decision will also follow these rank-order steps. Write your group decision on the Group Decision Sheet. Be ready to justify this decision to all the people affected by the policy you recommend.

PREDECISION TASK SHEET 1

The story includes a great deal of information about pesticides, what people eat, and the possible results of what is eaten. This Task Sheet will help you review this information. Write answers to the questions below. This task will prepare you to make a final decision as to what should be done about pesticide manufacturing, exportation, and use.

1. Official 5 used the phrase "circle of poison." In your own words, what is the "circle of poison"?

US->

2. According to the story, what are three actions the U.S. government is taking to reduce the amount of pesticide-carrying foods Americans eat?

   a.

   b.

   c.

3. In your own words, what is the "loophole" in our nation's laws that allows pesticide-carrying foods to be sold in America?

4. What are at least four statements in this story and the reports that reveal the use of science or technology?

   a.

   b.

   c.

   d.

5. What are at least three ways this "circle of poison" affects you?

   a.

   b.

   c.

6. What should the U.S. government policy be concerning the pesticides mentioned in this story?

7. Suppose you were living in one of the nations that was importing these pesticides because they were so inexpensive. Suppose you heard they had chemicals that were dangerous to human beings. In regard to future importation of these pesticides, what would your decision be?

8. In regard to America supporting companies that export to your nation pesticides forbidden in America, what would be your evaluation of Americans and the U.S. government?

9. Now that you know that certain American companies are selling pesticides they know will be used on foods eaten by Americans, what is your attitude toward these companies?

After you have completed your individual answers, share your responses with the others in your group. Add to, revise, or correct your responses as needed. Make sure everyone in your group comprehends this information before going on to the next page.

PREDECISION TASK SHEET 2

Consequential Analysis Chart

In the space below, write the possible or likely consequences of each of the proposed policies. List at least two good and bad consequences of each policy. In your list, include results that may happen very soon and those that may happen over a number of years.

Policy	Likely Consequences for the American People	Likely Consequences for Pesticide Makers	Likely Consequences for Foreign Nations
1. Urge Americans to wash their food better.	**a.** **b.**	**a.** **b.**	**a.** **b.**
2. Increase inspection of foods entering the U.S.	**a.** **b.**	**a.** **b.**	**a.** **b.**
3. Stop imports of foods from nations using these pesticides.	**a.** **b.**	**a.** **b.**	**a.** **b.**
4. Stop manufacture and export of these pesticides after 10 years.	**a.** **b.**	**a.** **b.**	**a.** **b.**
5. Stop manufacture and export of these pesticides immediately.	**a.** **b.**	**a.** **b.**	**a.** **b.**

After you have completed your individual answers for the chart, share your responses with your group. Add to, revise, or correct your responses as needed. Make sure everyone in your group comprehends this information before going on to the next page.

INDIVIDUAL DECISION SHEET

Record your personal decision about the policy the government should follow. On the space to the left of the five policies, place a 1 next to the option you most want the committee to accept. This is the policy you believe will work the best. Place a 2 next to the option you believe will work best *if and only if* the first one fails. Follow this procedure until all options have a rank number. Then answer the questions.

______ Allow continued exports of these pesticides and urge Americans to wash their foods more thoroughly.
______ Allow continued exports of these pesticides and increase inspection of all foods as they enter this nation.
______ Allow continued exports of these pesticides and stop all imports of foods from nations that use these chemicals.
______ Stop all manufacture and exports of these pesticides after 10 years to allow manufacturers to discover new pesticides.
______ Stop all manufacture and exports of these pesticides immediately.

1. The two most important reasons I selected policy _____ as the one that should be accepted are

a.

b.

2. The best thing that will happen should this policy be accepted is

3. The policy I ranked number 1 will affect me in the following three ways:

a.

b.

c.

4. This policy will affect people in foreign nations in the following two ways:
   a.

   b.

   c.

5. This policy will likely affect scientists in the following three ways:
   a.

   b.

   c.

6. The *worst* policy I could follow would be policy ______.

7. This is the worst possible policy for the following two reasons:
   a.

   b.

GROUP DECISION SHEET

Record your group decisions as to the policy the government should follow. As a group, agree upon one set of rankings for all the choices. Do not make a decision by majority vote. Find a choice that all members of your group will agree upon and support. On the space to the left of the five policies, place a 1 next to the one your group most wants the committee to accept. This is the policy your group believes will work the best. Place a 2 next to the option your group believes will work best *if and only if* the first one fails. Follow this procedure until all options have a rank number. Then answer the questions as a group.

______ Allow continued exports of these pesticides and urge Americans to wash their foods more thoroughly.

______ Allow continued exports of these pesticides and increase inspection of all foods as they enter this nation.

______ Allow continued exports of these pesticides and stop all imports of foods from nations that use these chemicals.

______ Stop all manufacture and exports of these pesticides after ten years to allow manufacturers to discover new pesticides.

______ Stop all manufacture and exports of these pesticides immediately.

1. The two most important reasons we selected policy ____ as the one that should be accepted are

a.

b.

2. The best thing that will happen should this policy be accepted is

3. The policy we ranked number 1 will affect us in the following three ways:
   a.

   b.

   c.

4. This policy will affect people in foreign nations in the following two ways:
   a.

   b.

5. This policy will likely affect scientists in the following three ways:
   a.

   b.

   c.

6. The worst policy we could follow would be policy ______.

7. This is the worst possible policy for the following two reasons:
   a.

   b.

REVIEW AND REFLECTION QUESTIONS

Suggested follow-up questions to focus and guide inquiry and learning.

a. As stated in this story, what is one of the major jobs of the EPA?

b. What are two of the duties of the Food and Drug Administration?

c. In your own words, what is the "circle of poison"?

d. What does it mean to *export* something?

e. What is a *pesticide?* What is their purpose?

f. What percentage of food you eat each day is likely to contain cancer-causing pesticides?

g. Suppose someone said Americans import their own poisons. According to this story, to what extent would this statement be true?

h. According to this story, how are Americans affected by the failure of the FDA to inspect imported foods?

i. There are many roles scientists have played and could play in this story about pesticides. What are at least three of these roles?

j. To what extent might American exports be poisoning people in other nations?

k. In what ways might scientists help solve the problem of pesticides being poisonous to people?

l. In what ways might scientists have contributed to the problem of pesticides being poisonous to people?

m. Suppose you washed food that was covered with pesticides in your kitchen sink. Where would the pesticides go?

n. Suppose the EPA did not adopt the policy of stopping production and exportation of these pesticides to foreign nations. What actions could you take as a citizen of this nation to oppose this action?

o. When scientists discover their efforts produce chemicals that are poisonous to people, what steps should they take next?

p. What are the worst consequences of the present EPA policy concerning the export of these pesticides?

q. Why would scientists support the sale of these pesticides to foreign nations?

r. These pesticides were found to be linked to cancer. What role might scientists have played in discovering this linkage?

s. Suppose a friend of yours was found to have cancer as a result of eating important foods with pesticides. In what ways would this news affect your reaction to the sale of these pesticides to foreign nations?

t. Many scientists are very aware that pesticides are causing harm to people both in this country and abroad, yet they still help develop these and other pesticides. What are your feelings toward these scientists?

u. When pesticides are used in these ways, are scientists or business people to blame for the deaths and health problems that result from the use of these pesticides?

v. Suppose you were an American farmer. What attitude might you have about importing foods to the U.S. that were grown with pesticides forbidden to be used in America?

w. Suppose you were a member of the government of a country that exports foods to the United States grown with these pesticides. If the U.S. government banned the import of food from your country, what would be your attitude toward the United States?

THOSE PESKY PESTICIDES

Wellington Paper Company is the largest producer of paper in the region. They own many acres of forests that supply them with wood pulp and other wood-related products. Eighty percent of the forest is hardwoods, such as oak, maple, and birch, used in construction and in making furniture. The other 20 percent is softwoods, such as pine, spruce, and fir, used to make paper and wood sculptures.

Reports showed that insects are infesting the leaves of young trees. The solution seemed simple. Orders were given to spray the forest. The pesticide killed the insects. However, after several months other effects of the pesticide were noticed. Trees that had been abundant had become scarce. Too late, company officials realized the steps they took to save the forest had destroyed it. Unless something was done, the company would lose millions of dollars.

After weeks of investigation, officials are presented with a number of steps they can take to replace now-scarce resources.

However, the company does not have surplus money to take all the steps at once. Officials are directed to consider which of the options should be tried first. If other moneys become available, the next option would be taken. One of the major problems is that some of the steps will take years before success or failure is known. Whichever option is selected first may be the only step the company ever takes to save itself.

Wellington Paper Company is considering these steps:

1. Buy large quantities of the scarce hardwoods from other lumber companies, Wellington's competitors. This purchase would create higher prices for Wellington's goods in the market place. High costs will likely result in fewer sales.
2. Spend money to find a chemical to eliminate the pesticides that have just killed the forest. This is risky because the new chemical, if found, might destroy even more forests. Even worse, there might be no chemical to stop the destruction by the pesticides.
3. Plant abundant softwood seedlings with the hope that enough of them would survive. This would increase supplies needed for paper. This is risky because the pesticides that killed the mature hardwoods are still in the soil.
4. Plant abundant hardwood seedlings with the hope that enough of them would survive. This would increase supplies needed for construction. This is risky because it takes 10 years for these seedlings to mature. Besides, the pesticides are still in the soil that killed the other hardwoods.
5. Buy acres of marginal land and try to develop it hoping hardwood will grow. This is risky because of the expensive irrigation and transportation expenditures that would be necessary.
6. Sell part of the company to a competitor to have enough money to stay in business. This is a risky proposition because of the uncertainty of controlling the future of the company.

Imagine you are one of the officials hired by Wellington Paper Company. You must help decide what steps the company should take. You also must decide what to do about the forest. Will the forest or the company be most important to you? Think about these things as you work through the following problem-solving steps:

- Review the story to this point. Define the problem, describe the efforts of Wellington to date, and consider what might happen to the forest if something is not done.
- Complete Predecision Task Sheets 1 and 2.
- Structure and record your personal final decisions on the Individual Decision Sheet. Then, structure and record your group's final decisions on the Group Decision Sheet.

PREDECISION TASK SHEET 1

On your own or as a member of a group, find out more information about pesticides. On this page, write down the ten most important pieces of information you want to remember about pesticides.

1.

2.

3.

4.

5.

6.

7.

8.

9.

10.

1. What are at least three important ways science is used in the production, use, and evaluation of pesticides?

a.

b.

c.

2. What are at least three important ways technology is involved in the production, use, and evaluation of pesticides?

   a.

   b.

   c.

3. What are at least five important ways your daily life would be affected by what companies such as the Wellington Paper Company decide to do about pesticide use?

   a.

   b.

   c.

   d.

   e.

After you have completed your individual answers for this sheet, share your responses with the others in your group. Add to, revise, or correct your responses as needed. Make sure everyone in your group comprehends this information before going on to the next page.

PREDECISION TASK SHEET 2

Working on your own, write down your answers to the questions below.

1. In your own words, what is a *pesticide?*

2. What are at least three important reasons humans use pesticides?
   a.

   b.

   c.

3. What led the lumber company to spray pesticides on the forest?

4. After the spraying, what specific things happened to the forest?

5. In the story, were the insects or the pesticides the biggest pest?

   What are two reasons that justify your answer to this question?
   a.

   b.

6. What does it mean for something to be *scarce?*

7. What is a *resource?*

8. In what three ways is it good that humans use pesticides to kill insects?
   a.

   b.

   c.

9. What might an ecologist say about using pesticides to preserve a forest?

10. If humans failed to use pesticides, what are two things that might happen to plant life on this planet?
    a.

    b.

After you have completed your individual answers for this sheet, share your responses with the others in your group. Add to, revise, or correct your responses as needed. Make sure everyone in your group comprehends this information before going on to the next page.

INDIVIDUAL DECISION SHEET

Wellington Paper Company has hired you to tell them the order in which their choices should be ranked. Rank the choices from the one you think is best (with a 1) to the one you think is worst (with a 6). The company asked you to keep in mind the image of the company and the immediate and long-term ecological effects. Your decisions will greatly affect what the Wellington Paper Company decides to do.

My Rank	Choices Available to the Company
______	Buy large quantities of the scarce hardwood from other companies.
______	Spend money to find a chemical to eliminate the pesticides.
______	Plant abundant softwood seedlings.
______	Plant abundant hardwood seedlings.
______	Buy and plant acres of marginal lands.
______	Sell part of the company.

1. My two strongest reasons for ranking option ______ as my first choice are
   a.

   b.

2. My two strongest reasons for ranking option ______ as my least favorite choice are
   a.

   b.

3. I think my first choice will have these four important results:
   a.

   b.

   c.

   d.

4. The two results I most want to avoid are
   a.

   b.

GROUP DECISION SHEET

Wellington has hired your group to tell them the order in which their choices should be ranked. The company asked you to keep in mind the company's image and the short- and long-term ecological effects. After individually ranking the choices, your group must rank order them in the Group Rank column. Without voting, the members of your group should agree on the rank for each choice. All members must agree on the the rankings of the options.

Group Rank	Choices Available to the Company
______	Buy large quantities of the scarce hardwood from other companies.
______	Spend money to find a chemical to eliminate the pesticides.
______	Plant abundant softwood seedlings.
______	Plant abundant hardwood seedlings.
______	Buy and plant acres of marginal lands.
______	Sell part of the company.

1. Our two strongest reasons for the selection of our first choice are

a.

b.

2. Our two strongest reasons for the selection of our least favorite choice are

a.

b.

3. We think our first choice will have these four important results:

a.

b.

c.

d.

4. The two results we most want to avoid are

a.

b.

REVIEW AND REFLECTION QUESTIONS

Suggested follow-up questions to focus and guide inquiry and learning.

a. Besides insects, what other life forms might be killed by pesticides?

b. What roles might scientists play in the lumber industry?

c. What roles might scientists play in the discovery and production of pesticides?

d. What roles might scientists play in efforts to limit the use of pesticides?

e. What are at least two positive things lumber companies do for the environment?

f. What are at least two negative effects on the environment of lumber company activities?

g. Some scientists work to discover and produce pesticides. Other scientists work against the use of pesticides. Why would scientists work against one another over this or other issues?

h. In the case of pesticides, would scientists be more concerned with using science or using the products of science?

i. To what extent is it good that scientists support products that may be harmful to life on this planet?

j. To what extent is it good that scientists work against the use of products that may be harmful to life on this planet?

k. Over 30 years ago Rachel Carson wrote a very readable book, *Silent Spring*. The book gave many examples of how chemicals used in pesticides and fertilizers were destroying our environment and poisoning humans. If these were actually happening and continue to happen, why would scientists continue to create new pesticides?

l. What are at least five products that humans have made from trees?

m. If products such as these pesticides had not been invented or developed by the use of science, how might your life be different from what it is today?

n. Suppose a friend or family member was found to have cancer as a result of pesticide poisoning in foods she or he ate. In what ways might your views of science and the use of science change?

o. Suppose you were driving a car, and an airplane spraying pesticides flew directly over your car, covering it with a light film of these pesticides. What would be your immediate response?

REPRESENTING THE PEOPLE

After years of being labeled a poor area, your district has begun to pay its own way. Newly established tourist businesses have attracted people from all over the nation. New jobs with good pay were created. New housing developments are planned. New stores and other businesses have opened. Unemployment is at an all-time low. In fact, people are moving in from other districts and states to work in your district. For the first time in decades, young people are talking about staying in the district rather than leaving it for better jobs elsewhere. The tourist industry has changed the entire district.

Because of the tourist attractions in the district, nearly 500,000 visitors come here each year and spend over $150 million at local businesses. Your district tourism council reports that 30 percent of their trade is made up of in-state and out-of-state people who drive to get to your district and return home the same day. In addition, an increase in over $18 million in local taxes was collected to pay for such public services as roads, schools, libraries, and police and fire protection. In many ways, your district is considered economically healthy.

You are a member of your state's House of Representatives. You were elected because of your efforts to bring "progress without destruction" to your district. You said bringing business into the area should not also mean bringing an end to the local environment. The local Environmental Action Group (EAG) supported your candidacy. Many farmers and older citizens voted for you. Your narrow victory was the result of a combined effort of EAG and non-EAG voters. The business people, whom you did not expect to support you, voted for your opponent. Since your election many more businesses have opened.

As the House session nears its end, there are five key bills yet to be passed. All five are very important to your district. If passed, they would likely do a great deal of harm to tourism in your district. Many businesses could close. Many people could lose their jobs. The district would once again become poor. However, the bills support the interests of the EAG. They would preserve the ecological balance and natural beauty of your district.

Mail and telegrams from the citizens of your district run three to one urging you to vote against all five bills. These messages indicate that if you vote against all five bills, you will not have problems getting reelected. However, you realize you don't have time to work to pass or defeat all five bills.

Members of the EAG remind you they helped get you elected. To vote against any of the five bills would be to turn your back on the protection of the environment and the EAG.

There is little time remaining in the session. All of the bills are given a 50–50 chance of passing. Some of the bills are designed to conserve energy. Some are designed to save the state money. Most have been proposed by the state EAG and the State Association of Farmers and Ranchers. The bills represent the strongest support environmental protection has ever received from the citizens of the state. You have to decide whether you will support the business and economic interests of your district over the interests of the environment. To date, you have successfully put off your answer to your constituents.

The five bills and brief descriptions of each are as follows:

1. *The Rhodes Bill (HB 95–109: Prevention of New Highways).* This bill would cut off all moneys to build new roads within the state. Money would be available only to repair and maintain

existing roads. The State Department of Transportation would dismiss all workers not needed to repair and maintain roads. If this bill passes, two new roads planned for your district will not be built. According to current estimates, these roads would bring nearly 350,000 tourists into your district the first year.

2. *The Petro Bill (HB 95–169: Increased Tax on Gasoline).* This bill would immediately add a 15-cents-per-gallon gasoline tax. It would give permission to the governor to extend the tax an additional 25 cents per gallon if conditions demand it. If this bill passes, gasoline prices will rise so high some people will not be able to afford the fuel to travel to the tourist attractions in your district.
3. *The Chips Bill (HB 95–214: Highway Patrol Expansion).* This bill would triple the size of the state's highway patrol department and purchase the necessary cars, radar, and other equipment. The enlarged force would effectively crack down on speed limit violators. Violators would pay a mandatory minimum fine of $100 to help pay for this patrol. If this bill passes, heavy enforcement of speed laws will take place throughout the state. Because the districts surrounding yours have not been able to enforce these laws in the past, these districts will receive extra patrols and heavier enforcement. Tourists may be scared off from traveling through these heavily patrolled districts.
4 *The Auto Tag Bill (HB 95–331: Increased Tax on Automobile License Tags).* This bill would require that automobile tags be sold to owners at a rate that includes miles-per-gallon and the amount of pollution. A car tag would cost a minimum of $225. Also, starting next year, all car owners would have to pay an additional fee for the miles their car had traveled during the year. If this bill passes, people may use their travel money to buy their auto tags. Since owners will be charged at the end of the year for all the miles they've driven, the bill will reduce the amount of extra driving they will do.
5. *The Emissions Bill (HB 95–332: Installation of Governors on Motor Vehicles).* This bill would require all in-state vehicles to be inspected each year for excessive air pollutants. The inspection would cost $25 per car. Drivers of both in-state and out-of-state cars with visible exhaust from tailpipes would be issued a $100 fine and be given 48 hours to repair the vehicle.

As you examined the five bills, you realized that if you voted for the bills, you would be voting to protect the state's environment. But you might also endanger the future of tourism in your district.

REPRESENTING THE PEOPLE

Earlier in the session, you decided to consider each bill as it came up. Then, one by one, you would make your decision as to which ones you wanted to see passed.

However, as you entered your office this morning, a group of business people from your district pay you a surprise visit—and bring seven reporters with them. They inform you they know the five bills may still pass. They know you have not yet decided for or against any of the bills. They report that if the bills pass, you will not be reelected. They demand to know which of the five bills you plan to spend the rest of the session trying to defeat. They demand an answer now!

As you prepare to respond to the business people, you are interrupted by a group of your district's Environmental Action supporters. They said they heard of this meeting and want to witness what is said. They brought along a television reporter to record the event.

Knowing you cannot spend all your efforts to defeat or pass all five bills, you announce you will work for or against the bills in order of their priority. With this announcement, the combined audience demands to know the order in which you will work on these five bills.

You must announce your decision.

In this situation, what actions will work best? What actions are you ready to support? How can you best represent the people in your district?

Before you make your final decisions, the Predecision Task Sheets will help you consider the options.

PREDECISION TASK SHEET 1

Before you consider the bills that may become laws in your state, take time to consider the situation. By yourself, develop and write answers to the questions below. Paraphrase the information from the story.

1. Suppose a person from another state visited your district before the tourist attractions were built. In describing the working and living conditions in your district, what are at least five statements he or she could have used?

   a.

   b.

   c.

   d.

   e.

2. Suppose that same person visited your district today. If asked to describe living and working conditions in your district, what would be at least five things he or she might say?

   a.

   b.

   c.

   d.

   e.

3. In your own words, what does it mean to "protect the environment"?

4. In your own words, what does it mean to "represent the people"?

5. As the representative in this story, what are your present beliefs about protecting the environment?

6. In the story, you are a member of your state's House of Representatives. What does it mean to be member of a state House of Representatives?

7. To help you make a decision about these bills, how might you use science?

8. To help you make a decision about these bills, how might you use technology?

After you have completed your individual answers for this sheet, share your responses with the others in your group. Add to, revise, or correct your responses as needed. Make sure everyone in your group comprehends this information before going on to the next page.

PREDECISION TASK SHEET 2

Before making your decision, consider each bill in light of its possible or anticipated consequences on the tourist trade and the local environment. In the chart below, describe each bill in your own words and state the possible consequences of each.

Bill Summary	Consequences for Tourist Trade	Consequences for Environmental Protection
1. The Rhodes Bill	**a.**	**a.**
	b.	**b.**
2. The Petro Bill	**a.**	**a.**
	b.	**b.**
3. The Chips Bill	**a.**	**a.**
	b.	**b.**
4. The Auto Tag Bill	**a.**	**a.**
	b.	**b.**
5. The Emissions Bill	**a.**	**a.**
	b.	**b.**

After you have completed your individual answers for this sheet, share your responses with the others in your group. Add to, revise, or correct your responses as needed. Make sure everyone in your group comprehends this information before going on to the next page.

INDIVIDUAL DECISION SHEET

Mark the bill that would be *most harmful to the tourist business* with a 1, the bill that would be next most harmful with a 2, and so on until you have marked the least harmful bill to your district's tourist business with a 5.

_____ The Rhodes Bill
_____ The Petro Bill
_____ The Chips Bill
_____ The Auto Tag Bill
_____ The Emissions Bill

1. I chose the ____________________ Bill as the *most harmful* to the tourist business for these important reasons:

a.

b.

c.

2. I chose the ____________________ Bill as the *least harmful* to the tourist business for these important reasons:

a.

b.

c.

Mark the bill that would be *most protective to the district's environment* with a 1, the bill that would be next most protective with a 2, and so on until you have marked the least protective bill to your district's environment with a 5.

_____ The Rhodes Bill
_____ The Petro Bill
_____ The Chips Bill
_____ The Auto Tag Bill
_____ The Emissions Bill

3. I chose the ______________________ Bill as the *most protective* to the tourist business for these important reasons:
   a.

   b.

   c.

4. I chose the ______________________ Bill as the *least protective* to the tourist business for these important reasons:
   a.

   b.

   c.

5. In my final decision, I chose to work toward passing/defeating the five bills in this order:

   ______ The Rhodes Bill
   ______ The Petro Bill
   ______ The Chips Bill
   ______ The Auto Tag Bill
   ______ The Emissions Bill

6. I chose to work toward passing/defeating these bills in this order for these important reasons:
   a.

   b.

   c.

7. If asked to explain why my decision best represents the people in my district, I would say

After you have completed the decision sheet by yourself, join with others in your group to reach a group consensus decision.

GROUP DECISION SHEET

Imagine you are now working as a committee of members of the House of Representatives. You are to make one set of decisions all members will agree upon and follow. This decision is to be made through consensus, not by majority rule. Mark the bill that would be *most harmful to the tourist business* with a 1, the bill that would be next most harmful with a 2, and so on until you have marked the least harmful bill to your district's tourist business with a 5.

______ The Rhodes Bill

______ The Petro Bill

______ The Chips Bill

______ The Auto Tag Bill

______ The Emissions Bill

1. We chose the ______________________ Bill as the *most harmful* to the tourist business for these important reasons:
   a.

   b.

   c.

2. We chose the ______________________ Bill as the *least harmful* to the tourist business for these important reasons:
   a.

   b.

   c.

Mark the bill that would be *most protective to the district's environment* with a 1, the bill that would be next most protective with a 2, and so on until you have marked the least protective bill to your district's environment with a 5.

______ The Rhodes Bill

______ The Petro Bill

______ The Chips Bill

______ The Auto Tag Bill

______ The Emissions Bill

3. We chose the ______________________ Bill as the *most protective* to the tourist business for these important reasons:
   a.

   b.

   c.

4. We chose the ______________________ Bill as the *least protective* to the tourist business for these important reasons:
   a.

   b.

   c.

5. In our final decision, we chose to work toward passing/defeating the five bills in this order:

   ______ The Rhodes Bill
   ______ The Petro Bill
   ______ The Chips Bill
   ______ The Auto Tag Bill
   ______ The Emissions Bill

6. We chose to work toward passing/defeating of these bills in this order for these important reasons:
   a.

   b.

   c.

7. If asked to explain why our decision best represents the people in our district, we would say

REVIEW AND REFLECTION QUESTIONS

Suggested follow-up questions to focus and guide inquiry and learning.

a. In the story, what two major groups attempted to influence your decision about which bills to support?

b. In the situation, how would the use of energy affect the prosperity of the tourist industry?

c. Suppose you had decided to rank order the five bills according to their environmental protection ability. Would you have rank ordered these bills any differently?

d. Individuals often make choices between what they want to do and what they ought to do. As the legislator in this exercise, when you had to decide between the people of your district and protecting your district's environment, did you feel confused?

e. What is the environment? Where is the environment?

f. What is the meaning of the phrase *environmental protection?*

g. In what ways can we protect the environment?

h. What is the opposite of *protect?*

i. In what ways can we damage the environment?

j. What have you done in the past month to damage the environment? To protect the environment?

k. Of these five bills, which would have the most affect on the way your family and you live?

l. To help protect the environment, what are at least five things you could begin doing today? What would keep you from doing these five things?

m. What is the job of an elected representative?

n. Why would scientists study things that would harm or protect the environment?

o. If a scientist reported that certain things harmed the environment, what would it take for you to stop doing the things that did this damage?

p. To what extent is it the job of the scientist to advise people what to do and not do to protect the environment?

q. To what extent is it your job to pay attention to the scientific evidence regarding ways to protect the environment?

r. To what degree are scientists required to follow their own advice?

s. If we fail to protect our environment, what are three things that could happen to the world in which we live?

t. How many tourists visit your town each year?

u. In what ways do tourists help your community?

v. In what ways do tourists harm the environment?

w. Based upon the evidence you have about the effect of visitors to your community, in what ways is it a good thing that these visitors continue to come to your town?

x. In this story, how did your final decision show that you represented the people in your district?

y. In this story, how might technology be used to protect the environment in this district?

z. How would each option in this story affect the use of technology in your district?

Negotiation Decision Episodes

Negotiation decision episodes provide students with a story in which the major character or group must make a decision from among listed options. This format requires students to consider the relative importance of all alternatives in situations in which they must give up something to gain, preserve, or protect other things they perceive as more important. Consequently, students develop the abilities to compromise, bargain, and negotiate—abilities they can use in many everyday situations.

OFF WE GO

Brownstone University has just received special funds for the commemoration of the first 30 years of space travel. This year five displays are to be selected and built for the university's science museum. The museum already has a variety of small space items such as space suits, lunar rocks, and small-scale models of rockets. These items may be used in the five larger displays that will be constructed. It has been suggested that life-size figures of humans, perhaps wax models, be included as part of the new displays.

It is hoped the museum will be a teaching museum. If the displays are well done, money will be provided for elementary and secondary school groups to visit the museum.

You are a member of the Display Selection Committee. Your committee will decide which five displays should be built to commemorate the first two decades of space exploration. The five best display ideas will be built this year. The next five might be funded and built in the coming year. The last five will probably never receive funding. In fact, your committee knows it will be giving up the last five space exploration displays to make sure the first five displays are built immediately.

The funding agency informed your committee that the selections must come from the list of space achievements below. No other events or things will be considered. Your committee will consider only these selections.

1. The first manufactured object to orbit Earth—*Sputnik I*—was launched by the Soviet Union in October 1957.
2. A dog named Laika was the first living creature placed in space. Laika was inside *Sputnik II,* launched by the Soviet Union in November 1957, for seven days.
3. The first satellite placed in orbit from Cape Canaveral, Florida, was the *Explorer I.* This was launched by the United States in January 1958.
4. The first person in space was cosmonaut Yuri A. Gagarin. He was carried into space by *Vostok I,* which was launched by the Soviet Union in April 1961.
5. The first U.S. astronaut to conduct a 15-minute, suborbital flight was Alan B. Shepard. The Mercury capsule, *Freedom 7,* was launched in May 1961.
6. The first U.S. astronaut to orbit Earth was John H. Glenn. He was inside the *Friendship 7,* launched in February 1962, which orbited Earth three times.
7. The first live transatlantic telecast from the United States to Europe was made possible when the satellite *Telstar I* was launched by the United States in July 1962.
8. The first planetary space craft, *Mariner II,* was launched in October 1962, by the United States. *Mariner II* traveled past Venus at a distance of 33,635 kilometers.
9. The Soviet Union's *Vostok 6* carried the first woman into space—cosmonaut Valentine Tereshkova—in June 1963.
10. The first person to walk in space was cosmonaut Alexei Leonov. He was carried into space by *Voskhod II,* which the Soviet Union launched in March 1965.
11. The first pictures relayed directly from the moon's surface were transmitted by *Luna 9.* The Soviet Union launched this satellite in February 1966.
12. The first people to walk on the moon were Neil Armstrong and Edwin Aldrin. These United States astronauts were carried to the moon in *Apollo II* in July 1969.

13. The first space station, *Skylab,* was launched by the United States in May 1973.
14. *Viking I* transmitted the first pictures from the rocky, red surface of Mars to the United States. This satellite was launched by the United States in August 1975.
15. *Voyagers I and II* left the United States in August and September 1979. *Voyager I* reached Jupiter in 1979 and collected data from Saturn in 1980. *Voyager II* reached Neptune in 1989. It was the first spacecraft to fly past three planets.

These are among the most important space events in the first three decades of space exploration. It will not be easy make a decision. To help you and the members of your committee reach the best possible decision, you should take time to consider the information and the space events. Remember, you are making this decision for the university and all the people who may visit the museum. In addition, you are revealing to the public what space events of those listed you believe are the most important ones to date.

Complete the Predecision Task Sheets. They are designed to help you make the best decision possible. Answer each sheet by yourself. Then the members of your group should share their answers with everyone else.

Once this sharing is over, make your final decision. After you have made your own choice, record your personal decision on the Individual Decision Sheet. Then meet with your group. Your group must make its own decision. Use the Group Decision Sheet to record your group's final decision.

PREDECISION TASK SHEET 1

Answer these questions individually before going on to the Decision Sheet.

1. What were at least three advancements in science that have been made *without* humans aboard the flights?
   a.
   b.
   c.
2. What were at least three advancements in science that have been made *with* humans aboard the flights?
   a.
   b.
   c.
3. Which three achievements would be considered the *most important* to scientists?
   a.
   b.
   c.
4. Which three achievements would be considered to be the *least important* to scientists?
   a.
   b.
   c.

5. Which three achievements would be considered to be the *most important* to society and people?
   a.

   b.

   c.

6. Which two achievements would be considered to be the *least important* to society and people?
   a.

   b.

7. What are three important reasons to use space probes to orbit Earth?
   a.

   b.

   c.

8. What are three important reasons to send space probes to or past other planets?
   a.

   b.

   c.

9. From 1957 to 1991, the United States and the former Soviet Union were the two major contributors to space exploration. What would be two sound reasons other countries did not get involved in space probes and exploration during this period?
   a.

   b.

After you have completed your individual answers, share your responses with the others in your group. Add to, revise, or correct your responses as needed. Make sure everyone in your group comprehends this information before going on to the next page.

PREDECISION TASK SHEET 2

Next to each space achievement, write two reasons it should be considered very important and two reasons it should not be. Doing this will make sure you consider each event from two different points of view. You may complete this task as a group.

Space Event	Two Reasons This Event Is a Very Important Space Achievement	Two Reasons This Event Is Not a Very Important Space Achievement
1. First human-built object to orbit Earth—*Sputnik I*—was launched in 1957.	**a.** **b.**	**a.** **b.**
2. A dog was placed in space for seven days in 1957.	**a.** **b.**	**a.** **b.**
3. *Explorer I,* the first U.S. satellite, was launched in 1958.	**a.** **b.**	**a.** **b.**
4. Cosmonaut Yuri A. Gagarin became the first person in space in 1961.	**a.** **b.**	**a.** **b.**
5. The first U.S. astronaut, Alan Shepard, went on a suborbital flight in 1961.	**a.** **b.**	**a.** **b.**
6. In 1962, John H. Glenn became the first U.S. astronaut to orbit Earth.	**a.** **b.**	**a.** **b.**
7. First live transatlantic telecast made by *Telstar I,* launched by the United States in 1962.	**a.** **b.**	**a.** **b.**

8. *Mariner II,* the first planetary spacecraft, traveled past Venus.	**a.** **b.**	**a.** **b.**
9. In 1963, Valentine Tereshkova became the first woman in space.	**a.** **b.**	**a.** **b.**
10. Alexei Leonov performed the first space walk in 1965.	**a.** **b.**	**a.** **b.**
11. *Luna 9* relayed the first pictures from the moon in 1966.	**a.** **b.**	**a.** **b.**
12. The first people to walk on the moon were Neil Armstrong and Edwin Aldrin, in 1969.	**a.** **b.**	**a.** **b.**
13. The first space station, *Skylab,* was launched by the U.S. in 1973.	**a.** **b.**	**a.** **b.**
14. *Viking I,* launched in 1975, transmitted the first pictures from Mars.	**a.** **b.**	**a.** **b.**
15. *Voyager I* reached Jupiter in 1979 and collected data from Saturn in 1980. *Voyager II* reached Neptune in 1989.	**a.** **b.**	**a.** **b.**

After you have completed your individual answers, share your responses with your group. Add to, revise, or correct your responses as needed. Make sure everyone in your group comprehends this information before going on to the next page.

INDIVIDUAL AND GROUP DECISION SHEET 1

Individually consider these fifteen advances in space exploration. Place a letter *T* in five blanks under the Personal Choice column to indicate your *top* choices for this year's space display. Place the letter *M* in five blanks to indicate your *middle* choices. Place the letter *B* in the five blanks to indicate your *bottom* choices. The *B* choices will never be included in the display.

Personal Choice	Group Choice	Space Achievements
______	______	**1.** First human-built space object, *Sputnik I,* orbits Earth.
______	______	**2.** First living traveler, Laika, a dog, sent into space.
______	______	**3.** First U.S. satellite, *Explorer I,* launched from Cape Canaveral, Florida.
______	______	**4.** First person, Yuri Gagarin, to travel in space.
______	______	**5.** First 15-minute suborbital flight by an American, Alan Shepard.
______	______	**6.** First American, John Glenn, to orbit the Earth.
______	______	**7.** *Telstar's* transatlantic telecast.
______	______	**8.** First planetary space craft, *Mariner II,* to travel past Venus.
______	______	**9.** First woman, Valentine Tereshkova, to travel in space.
______	______	**10.** First person, Alexei Leonov, to walk in space.
______	______	**11.** First pictures from moon's surface by *Luna 9.*
______	______	**12.** First people, Armstrong and Aldrin, to walk on moon.
______	______	**13.** First orbital space station, *Skylab.*
______	______	**14.** First pictures from surface of Mars by *Viking I.*
______	______	**15.** Planetary data from Jupiter, Saturn, and Uranus collected by *Voyagers I and II.*

Before going on to the next page, discuss the selections above with others in your group to reach a consensus agreement among the group members. Record your group choices under the Group Choice column as you did in the Personal Choice column. Then develop group answers to the questions on Group Decision Sheet 2.

GROUP DECISION SHEET 2

Now that your committee has selected the five displays to be built this year, you must justify your decisions. The questions below will help ensure that you develop a sound and thorough justification for your decisions.

1. In making our decision, the three most important things our committee tried to accomplish were

a.

b.

c.

2. The three strongest reasons for selecting our top five space events as the ones to be on display were

a.

b.

c.

3. These five are tied to science in the following four ways:

a.

b.

c.

d.

4. These five are tied to technology in the following four ways:
   a.

   b.

   c.

   d.

5. Our three reasons for thinking that our bottom group of five events were not as important as the first five are
   a.

   b.

   c.

6. We believe our society will benefit from this museum display of five space achievements in the following ways:
   a.

   b.

   c.

REVIEW AND REFLECTION QUESTIONS

Suggested follow-up questions to focus and guide inquiry and learning.

a. What is a *museum?*

b. What are the purposes of a museum?

c. What museums have you visited?

d. If you have visited museums, what kinds of things did you find in them?

e. In what ways might museums be connected to science?

f. Suppose a friend said that everything connected with space is also connected to science. In what ways would this statement be correct? Incorrect?

g. In what ways would a museum of space events also be a science museum?

h. You were asked to make a number of important decisions about space events. To what extent should nonscientists make decisions about what events are important to space exploration?

i. From the list of important space events given in this episode, which ones have had the least impact on your life?

j. Of the space events that were accomplished by the Soviet Union, which one do you wish the United States had done first?

k. Of the space events that were accomplished by the United States, which one do you wish the Soviet Union had done first?

l. Of these events, which one would you have most wanted to observe happening?

m. In what ways did science contribute to each of these accomplishments?

n. What kinds of scientists made each of these events possible?

o. In what ways do projects such as these contribute to science? To nonscientists?

p. What are three recent accomplishments of humans in their exploration of space?

q. What specific space events are missing from this list?

r. American astronauts and Soviet cosmonauts have died in their effort to explore space. To what extent is the exploration of space worth the lives of human beings?

s. The title of this episode is "Off We Go." How well does this title fit this particular activity?

t. When astronauts are launched into space, how much attention do you pay to their personal lives?

u. When astronauts are launched into space, how much attention do you pay to what they astronauts are trying to accomplish?

v. News reporters and some surveys indicate that today the vast majority of Americans do not pay much attention to most space missions. Why would Americans fail to pay attention to most space missions?

w. How much attention do you pay to current space exploration efforts?

x. What would it take to cause you to pay a lot of attention to space exploration efforts?

y. To what extent should people build museums to recognize past space accomplishments?

WHY? O WHY?? O WHY???

What killed the dinosaurs?

Scientists have long debated this question. According to most scientists, dinosaurs and flying reptiles suddenly vanished from the Earth about 65 million years ago. Some of the other animals that lived at the same time continue to live today. What could have caused the death of many dinosaur species without affecting thousands of other species?

Over the years scientists have developed a number of explanations or theories on what killed the dinosaurs. Theories are possible accurate explanations of what took place and why things take place as they do. *Theories* are not facts. They are ideas based upon the data available. Some theories seem to fit better than others.

As you read the theories that follow, remember that they are only possible ways to explain the extinction of the dinosaurs.

WHY? O WHY?? O WHY???

1. Disease Theory	According to this theory, an epidemic of some disease spread rapidly across the world. Dinosaurs, for some reason, had no protection against the disease and were wiped out. There are no accepted ideas as to what could have caused the disease to suddenly spread as it did. In addition, there is no evidence of disease in the fossil remains that have been found.
2. Egg-Eating Theory	Some scientists propose the dinosaurs became extinct because small mammals, including rats, ate all the dinosaur eggs before they hatched. It is believed there were too few mammals to do such a complete job of breaking all these eggs. There is no evidence of one species ever destroying all the eggs of another species.
3. Racial Senescence Theory	This theory says that each and every species has a certain life span. For a species, this life is similar to that of an individual. It has a start, a childhood, a youth, and an old age. Scientists who accept this theory believe extra armor, horns, and such are signs of the "old age" of a species. When such signs appear, the end of the race is near. Many fossils exists of dinosaurs with horns, extra armor, and so forth. It seems unlikely that every species of dinosaur would have died off at exactly the same time in this way.
4. Catastrophe Theory	This explanation says the Earth came very near colliding with a large object in space. This near collision caused all kinds of damage to the Earth. Large volcanoes, violent earthquakes, and huge tidal waves were some of the results. Many species were likely wiped out all at once. The large dinosaurs, unable to move rapidly or to find shelter, were the easiest targets for the destruction.
5. Survival of the Fittest Theory	Taken from the theory of evolution, this theory says that only the animals who are most able to survive continue to live. Those animals who cannot defend or save themselves will die. In the struggle for food, safety, and survival itself, the dinosaurs were not able to survive. What animal would have been more fit than every one of the thousand species of dinosaurs is not known.
6. Asteroid or Comet-Collision Theory	This more recent theory is growing in popularity. It says a large, rocky asteroid or small, icy comet entered the atmosphere and hit the Earth's surface. The effects of this collision included the immediate destruction of much of the land by earthquakes, volcanoes, and tidal waves. More damaging were the heating of the atmosphere and the dust layer that blocked sunlight from reaching the plants. Unable to produce oxygen for several years, the plants could not give the dinosaurs fresh oxygen to breathe. In addition, the dinosaurs had no escape from the heat and dust in the atmosphere.

These are only six of the theories scientists are suggesting. Of these six, which one seems the most likely reason for the deaths of all dinosaurs at the exact or near exact same time?

To complete this task, consider each theory very carefully. By yourself or with others in your class, try to find more information about each theory.

Then, select the two theories that you personally believe are the most likely explanations. Once you have made that decision, select the two that you believe are the most unlikely explanations. This will leave you with two other theories that you believe are not the best or worst explanations.

PREDECISION TASK SHEET 1

Below are questions for you to think about and answer. These questions are tied to the reading.

1. In your own words, what is a dinosaur?

2. According to the reading, a *theory* is a type of explanation. What else does the reading say a theory is?

3. How would theories about the death of dinosaurs be helpful for today's scientists?

4. Which of the theories is the most likely cause of this mass extinction?

5. What is it about this theory that makes you believe it is the best explanation?

6. Six theories were described in the reading. What are three other possible reasons dinosaurs became extinct?
   a.

   b.

   c.

7. If you were a scientist, what steps might you take to decide which theory is the most accurate?

8. Today's world has many endangered species. Humankind has already made many species extinct. One hundred years from now, what explanation will people use to explain the extinction of today's endangered species?

9. What are the two most important things the information on dinosaur extinction theories reveal to you about science or scientists?
   a.

   b.

After you have completed your individual answers, share your responses with the others in your group. Add to, revise, or correct your responses as needed. Make sure everyone in your group comprehends this information before going on to the next page.

PREDECISION TASK SHEET 2

Six theories on why the dinosaurs disappeared have been described. In the chart, write a short description of each theory. Then write at least two strong reasons to support each theory, and two reasons each theory could be inaccurate. If you can, write your own answers first, then add the reasons you read from your research on the theories.

Paraphrased Description	Two Strong Reasons to Believe This Theory Might Be Correct	Two Strong Reasons to Believe This Theory Might Be Incorrect
1. Disease Theory	a. b.	a. b.
2. Egg-Eating Theory	a. b.	a. b.
3. Racial Senescence Theory	a. b.	a. b.
4. Catastrophe Theory	a. b.	a. b.
5. Survival of the Fittest Theory	a. b.	a. b.
6. Asteroid or Comet Collision Theory	a. b.	a. b.

After you have completed your individual answers, share your responses with the others in your group. Add to, revise, or correct your responses as needed. Make sure everyone in your group comprehends this information before going on to the next page.

INDIVIDUAL DECISION SHEET

The six theories are listed below. Under the column labeled Personal Decision, place the letters MA on the line next to the two theories you consider to be the most accurate. Place the letters LA next to the two you consider to be the least accurate explanations. Leave the remaining two lines blank.

Personal Decision	Theory
______	Disease Theory
______	Egg-Eating Theory
______	Racial Senescence Theory
______	Catastrophe Theory
______	Survival of the Fittest Theory
______	Asteroid or Comet Collision Theory

1. What two major reasons make your two top choices the most likely to be *accurate explanations* for the extinction of the dinosaurs?
   a.

   b.

2. What two major reasons make your bottom two choices the most likely to be *inaccurate explanations?*
   a.

   b.

3. What are the two most important ideas this episode reveals to you about scientific theories?
   a.

   b.

GROUP DECISION SHEET

Imagine you are a member of a team of scientists trying to sort out the adequacy of the different theories. Make your decision by reaching a consensus rather than by using majority rule. Before you make your choices, answer this question: What are the characteristics of an adequate and acceptable theory? Place the letters *MA* next to the two theories you consider to be the *most accurate,* and place the letters *LA* next to the two you consider to be the *least accurate.* Leave the other two lines blank.

Group Decision	Theory
______	Disease Theory
______	Egg-Eating Theory
______	Racial Senescence Theory
______	Catastrophe Theory
______	Survival of the Fittest Theory
______	Asteroid or Comet Collision Theory

1. What two major reasons make your group's two top choices the most likely to be *accurate explanations* for the extinction of the dinosaurs?

a.

b.

2. What two major reasons make your group's bottom two choices the most likely to be *inaccurate explanations?*

a.

b.

3. What are the two most important ideas this episode reveals to your group about scientific theories?

a.

b.

REVIEW AND REFLECTION QUESTIONS

Suggested follow-up questions to focus and guide inquiry and learning.

a. In your own words, what is a *theory?*

b. When you use the word *dinosaur,* what kind of animals are you talking about?

c. If a typical dinosaur were standing outside the door, what would it look like?

d. How many million years ago did the dinosaurs suddenly die off?

e. What are the names of some dinosaurs that lived long ago?

f. What are some reasons scientists study dinosaurs?

g. What are some reasons scientists want to know what killed the dinosaurs?

h. This story was about theories on the death of dinosaurs. What other theories do you know?

i. Besides dinosaurs, what are the names of other animals that are now extinct?

j. Today, many animals and plants are on the endangered species list. What is an *endangered species?*

k. How does a scientist make up a theory?

l. To come up with a theory, what does a scientist have to do?

m. How might a scientist use a theory?

n. How might one theory be better than another?

o. Are you glad or sad that no dinosaurs may be alive today?

p. To what extent could a theory made up by a scientist ever be wrong?

q. Suppose someone discovered a live dinosaur. What should be done with it?

r. After reading all of these explanations for dinosaur extinction, which one do you find the most interesting? the most absurd?

s. Suppose a team of scientists just announced that none of the six theories above were accurate. Suppose they also announced that an entirely different situation led to the extinction of the dinosaurs. What would your feelings be toward the scientists who proposed these six theories? Toward the scientific theories?

t. Suppose you were a scientist. Suppose further your theory had just been found to be inadequate. How would you best describe your feelings?

u. Scientists claim you never "prove" a theory. Rather, you either find data to support or refute a theory. Why might they say you can't prove a theory?

v. Suppose someone said the information in this episode was more about history than science. In what ways might that person be correct? Incorrect?

w. To what extent is an historian a scientist?

x. What good is a theory?

y. In science, theories are tentative, meaning they are subject to change given new data. Because this is how scientists view theories, what important scientific theories of today may change?

EPISODE **16**

TOUGH DECISIONS

Your school's parent organization has always been active in raising money for various school projects. This year's money will go toward buying additional laboratory equipment. In the past there has been a lack of money for science equipment at Yonge School. *You are a student at Yonge School.* Your class has been asked to help select equipment needed for the science laboratory. You are excited to have funds just for science. You and your classmates realize the importance of wisely spending the money.

After looking through science catalogues, a group of parents in the organization marked the science equipment they might buy. The teachers are told the equipment must be selected from this list.

Your teacher and your class have been selected to choose from the list the eight *most important* items to purchase this year. If there is any money left over, you might get more items next year. Because of costs, there is a good chance your school will not get the eight pieces of equipment you mark as the least needed. In other words, the equipment you indicate as *least important* will probably never be bought. You decide to take a good look at the list of items for the science laboratory:

1. Microscopes with 50- and 100-power lenses, a mirror, and its own light would be used to study minute objects.
2. Dry cell batteries with an assortment of wires would be useful in energy and electrical experiments.
3. Test tubes, holders, beakers, and flasks would hold liquids when doing chemistry experiments or measuring levels of evaporation or temperature.
4. A movable model of the solar system would explain the location of objects in space.
5. Dissecting kits containing scissors, tongs, scalpels, files, and pins would better enable students to examine the internal organs of specimens.
6. Bunsen burners would be used to heat chemicals in order to observe changes in their form, color, state, or mixture.
7. A portable refracting telescope, which uses a series of lenses inside a long tube, would allow night and day observations of the sun, planets, and stars.
8. Pulleys, levers, strings, and ropes would help explain how people use tools to cut down on the amount of energy used and the work they have to do.
9. Small rocket kits, after being assembled, would demonstrate launching procedures and trajectory as well as the action-reaction principle of physics.
10. Camera kits would help students understand how the camera works as a type of a complex machine. Cameras could also be used to understand ideas about light, lenses, and sight.
11. Several terrariums, supplied with soil, plants, and/or animals, would allow students to observe changes within a limited space and environment.
12. Scales, weights, and rulers would allow measurements of all kinds of things to be made accurately.
13. Insect mounting boards would be an appropriate way to display insect collections.
14. Specimens for dissection would be used to study internal and external organs.
15. Several aquariums, equipped with fish, plant life, sand, and water, would give students a chance to take care of and to understand changes in living specimens.

16. Anatomy charts with labeled diagrams would make it easy to see and study the ten systems that make up the human body.
17. Small working models of a combustible engine would offer a way to better understand machines and the work they do.
18. Weather instruments, such as barometers, rain gauges, thermometers, an anemometer, and a wind vane, would be used in the study of patterns in climate and changes that take place in the atmosphere.
19. A supply of chemicals, such as acids, sulfur, mercury, and formaldehyde, would be needed to conduct experiments and preserve specimens.
20. Plankton nets would enable samples of microscopic plankton and algae to be collected from ponds, lakes, and streams.
21. A plastic model of a frog would give students a chance to study its many body parts and systems.
22. Large and small display cases would provide safe places to display rock, fossil, and insect collections.
23. A full-size human skeleton would give students a chance to name and know the location of bones in the human body.
24. Safety glasses, blankets, and fire extinguishers would be used to protect students and the school should an accident occur.
25. Crystal radio kits would help students know the mechanical parts of a radio and how radios change radio waves into sound.
26. Timers and stopwatches would be useful in accurately timing chemical reactions, the heating and cooling of materials, and the observable changes that occur within a specific time period.
27. Monthly star charts would point out the names and locations of constellations, stars, planets, and other space events that can be observed from the Earth.
28. Small cages, habitat material, and food supplies would enable students to observe the behaviors of various small land animals.

Now that you have read through and studied the descriptions of the types of science equipment, you need to take steps to make your final decisions. The Predecision and Decision Sheets that follow will guide your decision-making steps.

Imagine you are a member of this science class and you will help decide what equipment will be purchased. You are to make this decision as a group. But before your group decides, each of you will make your own personal decisions. Answer each of the Predecision Task Sheets by yourself before you meet as a group.

Before moving to Predecision Task Sheet, take a few minutes with your group to review the situation in the story to this point.

PREDECISION TASK SHEET (Optional)

From the descriptions in the story *and* other references if needed, write a definition for these terms:

1. microscope

2. dissect

3. formaldehyde

4. terrarium

5. telescope

6. anatomy

7. aquarium

These terms are mentioned in the story but are not defined. Referring to a dictionary or a science reference book, what is a clear definition for each term? State the definitions in your own words.

1. laboratory

2. constellation

3. specimen

4. anemometer

5. scalpel

6. trajectory

7. plankton

8. barometer

9. chemical

After you have completed your individual answers, share your responses with the others in your group. Add to, revise, or correct your responses as needed. Make sure everyone in your group comprehends this information before going on to the next page.

PREDECISION TASK SHEET 1

Across the top of this chart is a number of fields of science you might study in science classes at your school. Types of equipment that might be used in science classes is listed at the left. Consider each type of equipment and how it might be used. Place a check (✓) in the space below each field of science where that type of equipment might be used. Do this for each of the 28 types of equipment listed.

Fields of Inquiry

Equipment	Botany	Zoology	Solar System	Weather	Anatomy	Chemistry	Energy	Ecology	Geology	Machines	Physics
1. microscope											
2. dry cell batteries											
3. test tubes, beakers, flasks											
4. model of solar system											
5. dissecting kits											
6. Bunsen burners											
7. refracting telescope											
8. pulleys and levers											
9. small rocket kits											
10. camera kits											
11. terrarium											
12. scales and weights											
13. insect mounting boards											
14. dissection specimens											
15. aquarium											
16. anatomy charts											
17. model of combustible engine											
18. weather instruments											
19. chemicals											
20. plankton nets											
21. model of a frog											
22. display cases											
23. model of a skeleton											
24. safety equipment											
25. radio kits											
26. timers and stop watches											
27. star charts											
28. cages, water holders, food supplies											

After you have completed your individual answers, share your responses with the others in your group. Add to, revise, or correct your responses as needed. Make sure everyone in your group comprehends this information before going on to the next page.

PREDECISION TASK SHEET 2

Work on your own to complete this page.

1. I would describe a science laboratory in the following ways:

2. I believe a scientist uses a laboratory in these three ways:

a.

b.

c.

3. In what ways would this equipment be considered part of the tools a scientist uses?

4. Suppose a teacher said these tools or pieces of equipment had more to do with technology than science. What would be your response to this statement?

5. Suppose you had none of this equipment in your school. How could you learn to become a scientist with none of this equipment to use?

After you have completed your individual answers, share your responses with the others in your group. Add to, revise, or correct your responses as needed. Make sure everyone in your group comprehends this information before going on to the next page.

INDIVIDUAL DECISION SHEET

Working alone, consider these items that would improve the science laboratories in your school. Place the letters *MI* on the lines under the Individual Rating column next to the eight selections that would be the *most important* items to be bought this year. Place the letters *LI* on the eight lines that would be the *least important* items to buy for the lab. You will probably never have enough money to buy these items. Place an *M* on the twelve lines that are your *middle* choices. In case there is money left over from the first eight purchases, these middle items might be bought.

Individual Rating	Possible Science Equipment to Purchase
________	**1.** microscopes
________	**2.** dry cell batteries
________	**3.** test tubes, beakers, flasks
________	**4.** model of solar system
________	**5.** dissecting kits
________	**6.** Bunsen burners
________	**7.** refracting telescope
________	**8.** pulleys and levers
________	**9.** small rocket kits
________	**10.** camera kits
________	**11.** terrariums
________	**12.** scales and weights
________	**13.** insect mounting boards
________	**14.** dissection specimens
________	**15.** aquariums
________	**16.** anatomy charts
________	**17.** models of combustible engine
________	**18.** weather instruments
________	**19.** chemicals
________	**20.** plankton nets
________	**21.** model of frog
________	**22.** display cases
________	**23.** model of skeleton
________	**24.** safety equipment
________	**25.** radio kits
________	**26.** timers and stop watches
________	**27.** star charts
________	**28.** cages, habitat material, food supplies

1. The three most important reasons I selected the eight items of science equipment as the *most important* items to buy for the laboratory are
   a.

   b.

   c.

2. If my school obtained the eight *most important* pieces of equipment, I would be able to study these things:

3. If my school obtained the eight *most important* pieces of equipment, the best thing I could say about our laboratory and equipment is

4. The three most important reasons I selected the eight items of science equipment as the *least important* items to buy are
   a.

   b.

   c.

5. Without these eight pieces of science equipment, I will not be able to learn much about

6. Suppose a community says it values science and the training of future scientists but does not provide sufficient school tax money to pay for an adequate science lab. What would this say about how much the community is committed to teaching science?

GROUP DECISION SHEET

As a group, consider these items that would improve the science laboratories. Decide by consensus rather than by majority rule. Use group consensus for the questions that follow the rating. Place *MI* under the Group Rating column next to the eight selections that would be the *most important* items to be bought this year. Place *LI* on the eight that would be the *least important* items to buy. You will probably never have enough money to buy these. Place an *M* next to your twelve *middle choices*. If money is left over from the first eight purchases, these items might be bought.

Group Rating	Possible Science Equipment to Purchase
________	**1.** microscopes
________	**2.** dry cell batteries
________	**3.** test tubes, beakers, flasks
________	**4.** model of solar system
________	**5.** dissecting kits
________	**6.** Bunsen burners
________	**7.** refracting telescope
________	**8.** pulleys and levers
________	**9.** small rocket kits
________	**10.** camera kits
________	**11.** terrariums
________	**12.** scales and weights
________	**13.** insect mounting boards
________	**14.** dissection specimens
________	**15.** aquariums
________	**16.** anatomy charts
________	**17.** models of combustible engine
________	**18.** weather instruments
________	**19.** chemicals
________	**20.** plankton nets
________	**21.** model of frog
________	**22.** display cases
________	**23.** model of skeleton
________	**24.** safety equipment
________	**25.** radio kits
________	**26.** timers and stop watches
________	**27.** star charts
________	**28.** cages, habitat material, food supplies

1. The three *most important* reasons we selected the eight items of science equipment as the most important items to buy for the laboratory are

   a.

   b.

   c.

1. If our school obtained the eight *most important* pieces of equipment, we would be able to study these things:

3. If our school obtained the eight *most important* pieces of equipment, the best thing we could say about our laboratory and equipment is

4. The three most important reasons we selected the eight items of science equipment as the *least important* items to buy are

   a.

   b.

   c.

5. Without these eight pieces of science equipment, we will not be able to learn much about the following things:

6. Suppose a community says it values science and the training of future scientists but does not provide sufficient school tax money to pay for an adequate science lab. What would this say about how much the community is committed to teaching science?

Suggested follow-up questions to focus and guide inquiry and learning.

REVIEW AND REFLECTION QUESTIONS

a. What is a *laboratory?*

b. Where will you likely find a laboratory?

c. What words would most accurately describe a laboratory?

d. What is the purpose of a laboratory?

e. What types of people use laboratories?

f. What types of scientists use laboratories?

g. In schools, what purposes do laboratories serve?

h. What types of scientists would not use a laboratory?

i. What is the definition of *science equipment?*

j. To what extent can a person learn science without a laboratory?

k. To what extent can a person learn science without scientific equipment?

l. What items around where you live could you use as science equipment?

m. What things in the kitchen could be used to do science activities?

n. If you did not have access to real equipment, where else could you get information about a particular thing you were studying in science?

o. Dr. Frankenstein had a laboratory. In your own words, what was his laboratory like? In what ways was your image of Dr. Frankenstein's laboratory similar to the laboratory you have in your school? In what ways is your image of Dr. Frankenstein's laboratory different from the laboratory you have in your school?

p. In what ways could a library be a science laboratory?

q. What great scientific discoveries were probably made in laboratories?

r. Ben Franklin discovered that lightning was electricity during his famous experiment with a kite. During that experiment if he claimed "the sky was his laboratory," what would you have said to him?

s. If your school has no science laboratory, how much science could you learn?

t. How dangerous can a science laboratory be?

u. What is the purpose of having rules for proper conduct in a laboratory?

v. What rules have been established for proper behavior in your school's laboratory?

w. What are at least three reasons for following rules for safe laboratory behavior?

x. A great deal of information can be found in books and other reference sources found in libraries. In what ways can a library be a laboratory?

y. Suppose a parent claimed the school was buying technology and not science equipment. In what ways would this parent be correct? Incorrect?

z. In what ways is a piece of equipment a tool? In what ways is it technology?

WE'RE RUNNING OUT OF JUICE

Ms. Yirga, principal of Eustis Intermediate School, has received a notice from the school board: little money will be available to pay electricity and fuel bills for the rest of the school year. As one member of the board put it, "We cannot afford to operate all areas of the school. Quite simply, we're running out of 'juice'." Starting the following Monday, all schools in the district are to take steps to reduce their energy use drastically.

Ms. Yirga wants students to help her make the important decision as to what things would be done to cut the school's energy use. When you and your classmates came to school this morning, your teacher gave you a handout from listing nine things Ms. Yirga could do to cut energy use. The nine things are:

1. Stop using audiovisual equipment, including films, VCRs and monitors, overhead projectors, tape recorders, and computers.

2. Turn off air conditioning in the classrooms.
3. Turn off air conditioning in the library and computer room.
4. Allow warm water only for the showers in physical education.
5. Cancel all field trips.
6. Stop use of facilities, such as the gym and library, after 3:00 p.m.
7. Severely reduce the serving of hot and warm foods.
8. Close the reading center located in a portable classroom, which uses equipment, air conditioning, and lights.
9. In case of colder weather, turn on heaters only on extremely cold mornings, and turn them off at 10:00 a.m.

Ms. Yirga believes only some of these steps are necessary to cut down on the school's use of energy. So, three of these policies will be enforced immediately and three will not be done at all. The remaining three policies will be enforced if and when the first three fail to cut down the use of energy in the school.

You and your classmates are asked to consider these steps. In general, you all agree some of these rules are necessary. At the same time, you know all nine policies are not to be enforced immediately. Ms. Yirga is trying to enforce a school board policy. She could make her decision without asking your help. Your class is being asked to help decide which policies Ms. Yirga should enforce right away and which she should postpone unless absolutely necessary. To assist you and your classmates with your decision, Ms. Yirga asks that you do three things:

- Consider the good and bad things that could happen to students for each of the nine policies. In other words, if each policy was *not* selected, what positive things might this bring students? If each policy was selected, what bad things might this mean for students and the school?
- Select three policies that should be enforced immediately. State the reasons these policies should be followed. These are the things that will be done before anything else is done.
- Select three policies that should not be enforced at all, except in an extreme emergency. State the reasons these policies should not be enforced. These three areas are what the students are most willing to give up to get their top three choices.

The remaining three policies will be enforced only if enough energy is *not* saved by the first three policies.

On the Predecision Sheet, write your responses to Ms. Yirga's first request. Record your final decisions on the Decision Sheet. For each sheet, make your personal decisions first before working with members of your group.

PREDECISION TASK SHEET

Conservation Policy	Two Good Things That Would Occur If This Policy Is Not Enforced	Two Bad Things That Would Occur If This Policy Is Enforced
1. Stop using all audio visual equipment including films, VCRs and monitors, overhead projectors, tape recorders, and computers.	a. b.	a. b.
2. Turn off air conditioning in the classrooms.	a. b.	a. b.
3. Turn off air conditioning in the library and computer room.	a. b.	a. b.
4. Allow warm water only for showers in physical education.	a. b.	a. b.
5. Cancel all field trips.	a. b.	a. b.
6. Stop use of school gym and library after 3:00 p.m.	a. b.	a. b.

7. Severely reduce the serving of hot and warm foods in the cafeteria.	**a.** **b.**	**a.** **b.**
8. Close the reading center in the portable classroom, which uses lights, equipment, and air conditioning.	**a.** **b.**	**a.** **b.**
9. In case of colder weather, turn on heaters only on extremely cold mornings, and turn them off at 10:00 a.m.	**a.** **b.**	**a.** **b.**

After you have completed your individual answers, share your responses with the others in your group. Add to, revise, or correct your responses as needed. Make sure everyone in your group comprehends this information before going on to the next page.

DECISION SHEET

Individually consider the nine energy conservation policies. Place a *T* in the three blanks under the Individual Rank column that are your *top* choices to be enforced immediately. Place an *M* in the three blanks that are your *middle* choices or the next policies to be enforced. Place a *B* in the three blanks that are your *bottom* choices, those that should not be enforced unless absolutely necessary. After individually ranking the choices, meet with your group and rank the choices. Record these ranks under the Group Rank column as you did for the individual ranks. Without voting, the group members should agree on the rank for each choice.

Individual Rank	Group Rank	
______	______	**1.** Stop using all audiovisual equipment including film, VCRs and monitors, overhead projectors, tape recorders, and computers.
______	______	**2.** Turn off air conditioning in the classrooms.
______	______	**3.** Turn off air conditioning in the library and computer room.
______	______	**4.** Allow warm water only for showers in physical education.
______	______	**5.** Cancel all field trips.
______	______	**6.** Stop use of school facilities, such as the gym and library, after 3:00 p.m.
______	______	**7.** Severely reduce the serving of hot and warm foods in the cafeteria.
______	______	**8.** Close the reading center located in a portable classroom, which uses equipment, air conditioning, and lights.
______	______	**9.** In case of colder weather, turn on heaters only on extremely cold mornings, and turn them off at 10:00 a.m.

1. Our three reasons for believing the three policies marked *T* should be enforced at this time are

a.

b.

c.

2. Three ways we will be affected by these energy-saving moves are
   a.

   b.

   c.

3. Our three reasons for believing the three policies marked *B* should *not* be enforced unless absolutely necessary are
   a.

   b.

   c.

Persons participating in making this decision are

_______________ _______________

_______________ _______________

REVIEW AND REFLECTION QUESTIONS

Suggested follow-up questions to focus and guide inquiry and learning.

a. What policy did the school board expect Ms. Yirga to follow?

b. What caused the school board to pass such a policy?

c. How are the policies you were asked to consider related to problems of energy use?

d. What is the relationship between energy use and environmental protection?

e. What other ways of conserving energy might Ms. Yirga consider?

f. Is it fair for the school board to place the burden of the decision on principals?

g. If Ms. Yirga had made the decision by herself, what would be your feelings toward her?

h. If your own local school board began cutting energy use at your school, what would your feelings be toward them?

i. This episode concerned energy and the uses of energy. What does energy have to do with science?

j. In what areas of energy do scientists play a role?

k. Why would the study of energy be part of science?

l. To what extent is it good that scientists are involved in the study of energy and its uses?

m. In this story, how could scientists help students make the best decisions possible?

n. In what ways does energy conservation involve science?

o. How do scientists use energy?

p. How do you use energy?

q. How might you conserve more energy in your own life?

r. If we were to have another energy crisis, what steps would you take to conserve energy?

s. If we were to have another energy crisis, how would you expect scientists to act?

t. In what ways are scientists working now to find alternative energy sources?

u. If the price of all energy doubled overnight, what are five specific ways our society would be different next week from what it is today?

v. To what extent should a nation go to war to preserve its sources of energy from other nations?

w. Many people today waste a lot of energy. When you see them wasting energy, what are your feelings toward the waste? Toward the people?

x. What are four reasons a person such as you would waste energy?

y. What impact does the use of particular types of energy have on the environment?

z. In this story, one member of the school board used the word *juice* instead of electricity. Why do some people use the word *juice* to refer to electrical energy?

SCIENTISTS—THEY'RE EVERYWHERE!

Pretend you will be given one week to select the three best types of scientists to talk to your class concerning the future of all people. After all, the scientific studies and research done today will certainly affect your future and the future of the planet. As scientists of the past have contributed to today, today's scientists will impact the events and discoveries of tomorrow. This is how it has been for years and years.

Before you decide which scientists to invite to your class, review some of the important things you know about science, the work scientists do, and the methods scientists follow in their investigations. Without knowing these things, you would not be able to fully comprehend the work and abilities of scientists.

SCIENTISTS—THEY'RE EVERYWHERE!

Pretend your research helped you in making the following outline about science, scientists, and one method of science:

Four of the ways *science* has been defined are

- A set of steps followed to obtain and verify facts, answers, and explanations.
- The asking and answering of questions about any thing or event in the universe.
- A collection of organized, systematically researched facts and explanations about naturally occurring objects or events.
- A way of thinking or an attitude about inquiry, research, and investigations. A state of mind or viewpoint a person accepts.

If you were a *scientist,* you would be expected to

- Report events and data accurately and honestly
- Maintain heavy skepticism toward apparent facts
- Refuse to jump to quick conclusions
- Follow organized steps in searching for facts
- Question existing facts, events, and their explanations
- Accept the probability of change in present conceptions, theories, knowledge
- Be open-minded toward new ideas
- Search for mistakes and omissions
- Admit your own mistakes
- Accept the obligation to test your own conclusions
- Encourage others to test your conclusions

A scientific method is a series of steps some scientists use to establish scientific laws, principles, and explanations. They follow these steps to guide their investigations and inquiry. There are a number of scientific methods of inquiry and investigation scientists use. The steps and substeps of *one of the many methods* scientists use are given below.

STEPS OF ONE METHOD OF SCIENTIFIC INVESTIGATION

A. Make observations and then determine what problem exists or what question needs to be answered.
   - Study known facts
   - Gather other facts from personal experiences
   - Analyze information
   - Determine the problem or unanswered question

B. Develop a hypothesis (prediction or guess) about one or more possible solutions to the problem or answers to the question.
   - Propose a possible correct explanation to the problem
   - Propose a possible workable solution to the problem
   - Propose a possible correct answer to the question

C. Conduct experiments to support or reject the hypothesis or a possible answer.

D. Collect and analyze data from experiments.

E. Allow others to test the hypothesis.

F. Draw conclusions about and evaluate each hypothesis or possible answer.
   - If the hypothesis is supported by the data, the principle or explanation is preserved and used.
   - If the hypothesis is not supported by the data, the scientist tries other reasonable predictions about the same problem.
   - If the proposed answer is supported by the data, it is accepted as a tentative answer under those circumstances.

With this information in mind, you begin reading job descriptions for various scientists. These scientists must be able to observe, hypothesize, conduct experiments, collect data, and make honest evaluations to help ensure our future in the universe. You realize this may take months, years, or a lifetime. You are surprised by the variety of scientists you find. The following are the nine job descriptions you were able to find in one week:

1. *Oceanographer.* Oceanographers study the ocean and its floor. Oceanographers are interested in tides, waves, currents, temperatures, and minerals in the ocean. They also investigate the mountain ranges, volcanoes, canyons, trenches, and level plains in and under the ocean.
2. *Meteorologist.* Meteorologists study the atmosphere, weather, climate, and weather forecasting. They are interested in changes in temperature, wind direction, and speed, pressure and moisture in the air, clouds, and precipitation (rain, snow, sleet, hail).

3. *Historian.* Historians study the social, political, and economical changes taking place over periods of time and space. They are interested in reporting public and private facts about those events and people. For example, a historian would study ancient Egypt, trends in music in a particular period, or the life of a past president.
4. *Geologist.* Geologists study the structure and the history of the earth: how the layers of the earth are formed and how they will probably change. They are interested in caves, glaciers, fossils, volcanoes, earthquakes, rock samples, and mines.
5. *Chemist.* Chemists study the matter (anything except space) that makes up the living and nonliving parts of the earth. Chemists are interested in what matter is made of and how those materials change. In laboratories, they study reactions of chemicals, such as iron, nickel, and uranium, to other materials. In or out of a laboratory, a chemist might analyze blood samples, particles of fabric, or samples of polluted water.
6. *Biologist.* Biologists study all living things that make up the earth. There are two kinds of biologists. A botanist studies plants; a zoologist studies animals. Biologists are interested in the causes and cures of diseases, hereditary characteristics, the environment, and the anatomy of organisms.
7. *Astronomer.* Astronomers gather and verify data and explanations about heavenly bodies and the space that lies between them. They are interested in the planets, the sun, natural satellites, asteroids, meteors, and stars. Astronomers make use of telescopes and other instruments to study heavenly bodies.
8. *Archaeologist.* Archaeologists study societies and cultures of ancient times by looking at tools and other objects left behind by early societies. Archaeologists are interested in arrowheads, pottery, and other objects (artifacts) from other cultures. They also study remains of ancient monuments, temples, and palaces.
9. *Anthropologist.* Anthropologists study the connections between people, their origins (beginning), and their ways of life. They are interested in physical, social, and environmental characteristics among different groups of people. For example, an anthropologist would study a primitive society in Australia.

Before selecting the three "best" scientific occupations you want your class to hear about, individually complete the Predecision Task Sheets. Once you have completed each Predecision Task Sheet, share your answers with others in the class. Make sure all members of your group comprehend the ideas on each before moving to the next one. Then make your final decisions on the Decision Sheet.

PREDECISION TASK SHEET 1

Working on your own, answer the questions below to help you interpret the information in the story so far. Do not copy the information word for word from the story. Write paraphrased answers to these questions.

1. In your own words, what are three of the definitions of *science?*

a.

b.

c.

2. At the present time, what definition of science do you like the best?

3. In your own words, what are at least six things scientists are expected to do?

a.

b.

c.

d.

e.

f.

4. Of the expectations listed, which three would be the hardest for you to do?

a.

b.

c.

5. Many teachers and science textbooks act as though there was only one scientific method. There are actually many methods of conducting sound scientific inquiry. How could there be more than one methods of conducting scientific inquiry?

6. If your own words, what are the steps to complete as one of the methods of scientific inquiry?

After you have completed your individual answers, share your responses with the others in your group. Add to, revise, or correct your responses as needed. Make sure everyone in your group comprehends this information before going on to the next page.

PREDECISION TASK SHEET 2

Complete the chart. Remember, scientists invent interpretations and hypotheses based on their observations. They do research to support, challenge, or reject hypotheses, interpretations, and facts. Some collect and study data from experiments, while others cannot conduct experiments at all. Write a brief description of how each scientist might use the scientific method outlined in the story.

Scientist	People in This Area Are Likely to Use a Scientific Method to Study the Following Things
1. Oceanographer	
2. Meteorologist	
3. Historian	
4. Geologist	
5. Chemist	
6. Biologist	
7. Astronomer	
8. Archaeologist	
9. Anthropologist	

After you complete your individual answers, share your responses with your group. Add to, revise, or correct your responses as needed. Make sure everyone in your group comprehends this information before going on to the next page. Make sure you can describe how each of these occupations represents science.

PREDECISION TASK SHEET 3

Following the example for the oceanographer, complete the chart. Describe things each of the scientists studies then write how each area of science would help make a better future.

Scientist	Area of Study for This Scientist	This Importance of This Occupation for the Future
1. Oceanographer	Explores what lives and exists in the ocean.	Finds ways to use and protect the ocean and its resources.
2. Meteorologist		
3. Historian		
4. Geologist		
5. Chemist		
6. Biologist		
7. Astronomer		
8. Archaeologist		
9. Anthropologist		

After you complete your individual answers, share your responses with your group. Add to, revise, or correct your responses as needed. Make sure everyone in your group comprehends this information before going on to the next page. Make sure you can describe how each of these occupations represents science.

DECISION SHEET

Working on your own, place the letter *T* in the three blanks under Individual Choices column to mark the *top* kind of scientist you want to have talk to your class about the future of all people. Place the letter *M* in three blanks to indicate your *middle* choices in case there were enough interest in your class to invite six scientists during the year. Place a *B* in three blanks to indicate your *bottom* choices, which would never be invited to speak to your class. After you've made your own choices, share your choices with members of your group to review these nine descriptions of scientific jobs with you. As a group, decide what would be your top, middle, and bottom choices. Mark your group decisions under the Group Choices column.

Group Choices	Individual Choices	Scientist
______	______	**1.** Oceanographer
______	______	**2.** Meteorologist
______	______	**3.** Historian
______	______	**4.** Geologist
______	______	**5.** Chemist
______	______	**6.** Biologist
______	______	**7.** Astronomer
______	______	**8.** Archaeologist
______	______	**9.** Anthropologist

1. The most important reason why I want my *top* three choices to speak is

2. These three types of scientists could help the future of all people in the following two ways:

a.

b.

3. The reason why I picked the three *bottom* occupations last is because

REVIEW AND REFLECTION QUESTIONS

Suggested follow-up questions to focus and guide inquiry and learning.

a. According to this episode, what is *science?*

b. Of the three definitions given, which one would nearly all scientists accept as being the most accurate?

c. What does a scientist do?

d. In what ways would a historian be considered a scientist?

e. If you used the method of science described above to study history, how would you go about studying an important event of the past?

f. Before this episode, would you have included historians as scientists?

g. What are the names of other occupations of science that were not listed here?

h. Of all of the things scientists think and do, which determine that they are really scientists?

i. How do these different occupations contribute to improving our world or way of life? Harming our world or way of life? Of these occupations, which would you most want to join? Which would you most want to avoid?

j. The title of this episode is "Scientists—They're Everywhere!" In what ways did the list of occupations prove that scientists really are everywhere?

k. In what ways are you a scientist?

l. If you do not consider yourself a scientist, what would you have to do to become a scientist?

m. If you were asked to find scientists in your community, where would you go to find them?

n. In what occupations would you *not* find scientists working?

o. Suppose you memorized your science book. If you could recall all the facts in this book, would you be a scientist?

p. Most scientists say science is a way of thinking. What do they probably mean by this?

q. How hard would it be for you to think like a scientist?

r. For what reasons might a person claim women and men may both become excellent scientists?

s. Research has revealed that both girls and boys can become excellent scientists and that neither sex is naturally a better scientist than the other. How do these research findings compare with your ideas concerning males and females as scientists?

t. What makes a scientist an excellent scientist?

u. Who are some of the scientists you admire?

v. What things would prevent you from becoming an excellent scientist?

Invention Decision Episodes

Invention decision episodes provide students with a story in which the major character or group must make a decision, but is free to make any decision that is consistent with the given situation. During these learning tasks, students are encouraged to select or invent an appropriate decision to respond to and resolve the situation before them. In some episodes they may select from among any of the given options, combine them to form new choices, or reject them all. This episode format helps prepare students for decision-making situations in which they have a great deal of flexibility in what decisions they can make.

EPISODE **19**

BO OR ZO?

Patsy and Goodie both want to become scientists. They agree that biology, the study of all living things, is the area they are most interested in understanding.

Patsy thinks that to be a real scientist, a person must study the area of biology called *botany*. Botanists study plants. Goodie says a real scientist is a person who studies the other branch of biology, *zoology*. Zoologists study animals.

“It is important for zoologists to know the ways in which animals are alike and different,” says Goodie. “All the animals we know of that have ever lived have been grouped into classes or categories according to their common characteristics. This classification system is called a *taxonomy*. With this system zoologists can even compare today’s animals with those of millions of years ago. A taxonomist in zoology plays an important role in biology.”

“I disagree,” says Patsy. “A taxonomist in botany must arrange plants in classes or categories ranging from the most primitive to the most complex. Just as in the animal world, plants are arranged in these classes on the basis of their common characteristics. And, again as in the animal world, each plant is given a scientific name that's the same throughout the world."

"Yes, but one exciting way a taxonomist uses the classification system in zoology is to work with fossils," remarks Goodie.

"A botanist also studies fossils," interrupts Patsy.

"I know," says Goodie, "but I think the category system in zoology is more important. I like studying extinct animals like the dinosaurs. We can use animal taxonomy to classify past and present animal life."

"If you are looking at the past, a geneticist in botany is another important area of specialization. Geneticists study how plants transmit hereditary characteristics. They may also discuss ways to produce new kinds of plants by crossbreeding."

"Geneticists in zoology are also interested in the beginnings of animals," Goodie quickly adds. "They study how each feature is passed from parents to offspring. For instance, a geneticist in zoology can tell you why you have blue eyes and not brown ones."

"OK, so maybe a geneticist in biology and zoology do similar jobs related to their area of specialization. But I know a real difference," declares Patsy.

"It couldn't be that different."

"Do you know what a cytologist does?" asks Patsy.

"A cytologist?" questions Goodie.

"Yes. A cytologist studies the cell. The protoplasm surrounds the nucleus of a cell. Remember? The nucleus controls all of the activity of the cell. A cytologist, with the help of a microscope, can see things like the green material in plants called chlorophyll. This study is important to understanding all living plant cells. Besides, if plants didn't give off oxygen, animals couldn't live."

Goodie quickly defends the zoologists. "A part of the study of anatomy in zoology deals with the study of cells. Anatomy is the study of all the structures and systems of animals. Humans, as animals, are part of this study."

"What happens when some of those cells become diseased?" asks Patsy.

"In zoology, pathologists will study the diseased cells and tissues. They will try to find the cause and cure of the disease."

Patsy anxiously continues. "It's the same thing in botany. A pathologist in botany knows that the cause of the disease may be from bacteria, viruses, rusts, or molds. Sometimes, when they know the life cycle of a particular plant, plant pathologists can help to prevent or stop the disease."

"An animal pathologist is also interested in finding a cure for the disease," says Goodie.

"It is important for an ecologist in botany to know how plants are affected by temperature, soil, humidity, and sunlight. All plants are part of a community. An ecologist knows how a plant fits into that community."

"Could plant ecologists predict what might have happened to an area of barren land?" asks Goodie.

"Sure they could," replies Patsy. "They should be able to make a prediction like that in any part of the country. They should also be able to predict many of the kinds of animals that will probably live in each area."

"Looks like we have another similar area of specialization," declares Goodie. "An ecologist in zoology understands and studies how animals get along in certain environments."

"In all environments?"

"Well," begins Goodie slowly, "there might be one environment that would be different. In zoology, a paleontologist studies things that used to be alive but that are now extinct. So you might say the geological fossil environment would be an area of specialization."

"But examples of extinct plants can also be found in fossils," remarks Patsy. "I wonder if there are more fossilized plants or animals."

"I don't know, but it seems we are back to our previous discussion of fossils. Could that be a key in understanding which is a real scientist—a zoologist or a botanist?"

Imagine you are a teacher. Imagine Patsy and Goodie came to you asking these questions:

- What does a real scientist do?
- What makes botany and zoology fields of science?
- Which branch of biology—zoology or botany—might be considered more a field of science than the other?
- Might botany or zoology be considered more of a science than the other branch?

From the conversation you just read and what you already know about science, what answers would you give Patsy and Goodie? Complete the Predecision Task Sheets on the following pages before you answer their questions.

PREDECISION TASK SHEET 1

Listed below are terms used in the story. Write a definition for each based on the conversation you just read and what you already know about science.

1. A zoologist is
2. A botanist is
3. A pathologist is
4. A taxonomist is
5. A geneticist is
6. Anatomy is
7. Crossbreeding is
8. A cytologist is
9. A nucleus is
10. Bacteria is
11. A fossil is
12. A scientist is

Now go back and check your definitions with the descriptions in the story. If you need more help in defining the terms, consult a science or other reference book.

After you have completed your individual answers, share your responses with the others in your group. Add to, revise, or correct your responses as needed. Make sure everyone in your group comprehends this information before going on to the next page.

PREDECISION TASK SHEET 2

Before working on the chart below, complete this statement:

My description of a scientist is

Write the scientific name for the job described under Area of Study. Place a check on the blank if a botanist and/or a zoologist would study the item listed.

Scientific Name	Area of Study	Botanist	Zoologist
______________	1. disease	______	______
______________	2. heredity	______	______
______________	3. cells	______	______
______________	4. bacteria	______	______
______________	5. fossils	______	______
______________	6. sunlight	______	______
______________	7. rust	______	______
______________	8. classification	______	______
______________	9. skeleton	______	______
______________	10. chlorophyll	______	______
______________	11. barren land	______	______
______________	12. dinosaurs	______	______

After you have completed your individual answers, share your responses with the others in your group. Add to, revise, or correct your responses as needed. Make sure everyone in your group comprehends this information before going on to the next page.

INDIVIDUAL DECISION SHEET

1. Suppose you were Patsy. What are two specific reasons you can give to prove a botanist is a scientist?
   a.

   b.

2. Suppose you were Goodie. What are two specific reasons you can give to prove a zoologist is a scientist?
   a.

   b.

3. Suppose someone said, "Both branches of biology are equally important." How would you react to this statement?

4. Continue to imagine you are a teacher of Patsy and Goodie. They asked you for advice about whether zoologists and botanists are both real scientists. What are four of the most important things you would tell Patsy and Goodie?
   a.

   b.

   c.

   d.

5. In answering the first question Patsy and Goodie asked, what are four important things that a real scientist does?
   a.

   b.

   c.

   d.

6. Which branch of biology would you consider to be more a field of science than the other? ______________________________

   What are your two best reasons for this decision?
   a.

   b.

7. What are two important contributions botanists and zoologists make to your life and the way you live?
   a.

   b.

8. In doing their work as scientists, what are two important ways botanists and zoologists use technology?
   a.

   b.

GROUP DECISION SHEET

Imagine you are a group of science teachers and students working together to respond to Patsy and Goodie's questions. Write answers the questions below. The answers should represent the best thinking of all members of your group. They also should be ones all group members agree on and support. Make your group decisions by consensus, not by majority rule.

1. What are the two best reasons to verify that a botanist is a scientist?

a.

b.

2. What are the two best reasons to verify that a zoologist is a scientist?

a.

b.

3. Suppose someone said, "Both branches of biology are equally important." How would your group react to this statement?

4. Continue to imagine that Patsy and Goodie asked you for advice about whether zoologists and botanists are both real scientists. What are the four most important things you would tell them?

a.

b.

c.

d.

5. In answering the first question Patsy and Goodie asked, what are four important things that a scientist does?
   a.

   b.

   c.

   d.

6. Which branch of biology might be considered the most important branch?

   What are your two *best* reasons for this decision?
   a.

   b.

7. What are two important contributions of botanists and zoologists to your life and the way you live?
   a.

   b.

8. In doing their work as scientists, what are two of the most important ways botanists and zoologists use technology?
   a.

   b.

REVIEW AND REFLECTION QUESTIONS

Suggested follow-up questions to focus and guide thinking and learning.

a. In your own words, what is *botany?*

b. What does the prefix *bio-* mean?

c. What does the suffix *-ology* mean?

d. Of the areas of botany and zoology mentioned in this episode, which area studies fossils?

e. By asking so many questions, would you consider Patsy and Goodie to be inquirers? To be scientists?

f. What are at least two ways botany is different from zoology?

g. Suppose you learned information about living things only from books, films, and television shows. In what ways would you be a scientist? In what ways would you not be a scientist?

h. In what places would a botanist do his or her work?

i. What living things have you tried to study?

j. Biologists study living things or things that lived at some other period of time. What is it that all living things have in common?

k. What is a *living thing?*

l. What are at least three reasons people would want to study living things?

m. How could we use information about living things?

n. Is the study of living things more or less important than the study of nonliving things?

o. What living things would you prefer to study?

p. To what extent are you a botanist? A zoologist?

q. Suppose scientists didn't study living things. With no such information, how would our world today be different?

r. What are at least three benefits to our society that result from our study of living things?

s. Suppose you had a million dollars to give either toward the study of plants or the study of animals. To which would you give this money to help scientists do more research?

t. When you do research, to what extent are you a scientist?

EPISODE **20**

PULLING TEETH

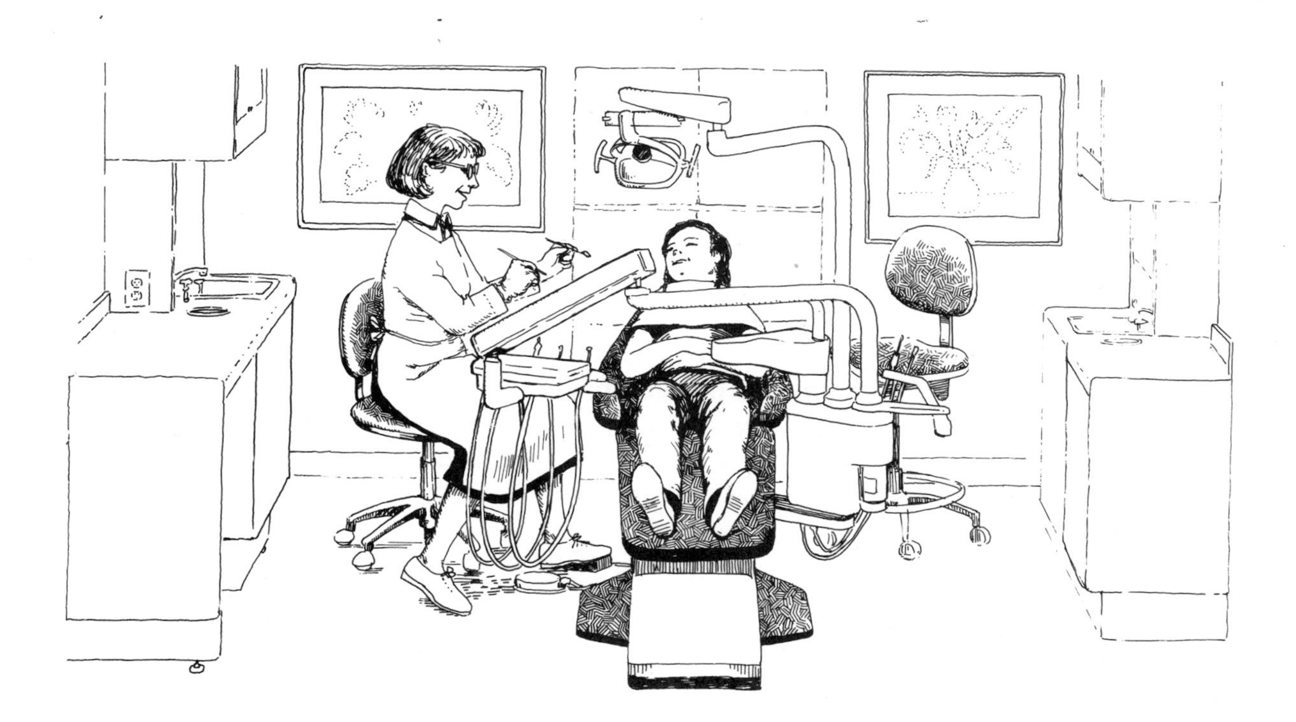

Mercury's danger to humans is well known. There is some increasing evidence that mercury in people's silver dental fillings endangers their health. Silver amalgam fillings, believed to be in the mouths of over a 100 million people, are composed of about 50 percent mercury. They also contain about 35 percent silver, 15 percent tin or tin and copper, and a trace of zinc. The U.S. Food and Drug Administration approved the use of mercury in 1976.

Small amounts of mercury from these fillings get into the bodies of nearly everyone who has them. Because of the damage mercury can do to the body, some scientists recommend that dentists stop all use of silver amalgam fillings. In addition, it is proposed that the public be informed that those who have such fillings should consider having them replaced. However, most members of the dental profession claim there is no danger to human health from amalgam because it is safe to use. They claim there is no research evidence proving amalgam is unsafe for human use.

PULLING TEETH

A panel has been formed by the City Health Commission to make a decision that could affect thousands of people in your community. It will also affect how dentists treat tooth decay and cavities. *You are a member of that panel.* Consider the evidence, and then recommend a course of action. In the past, the recommendations of such special panels have been approved without changes by the commission. For all practical purposes, you will be making the policy that will be enforced in your community.

For several days witnesses have testified before your panel. A summary of their testimony follows.

Witness A: A Canadian dental research scientist
"Research suggests mercury from amalgam fillings causes kidney malfunction in test animals—in one case, sheep. One problem with this major study is that sheep wear down their teeth much faster than humans do. Many studies have shown that amalgam increases the level of mercury in the blood and that mercury from fillings can lead to malfunctions in body organs. More recent information indicates that mercury from fillings passes through the placentas of human mothers and concentrates in the brains of fetuses."

Witness B: A representative of the American Dental Association
"We believe that dental amalgam is still a safe, effective restorative material. Hence it poses little to no danger to the health of humans. There is no reason for the scientific community or public to be alarmed or disturbed by hearsay about the dangers of amalgam fillings. There is no reason for people to seek the removal of such fillings."

Witness C: A dentist and researcher at a California university
"Patients should be worried about these fillings and the health risks involved. I encourage all of my patients and students to discuss alternative methods when I find a new cavity. I also advise them to consider other materials when an old filling needs to be replaced. However, people, especially pregnant women, should not seek large-scale removal of their fillings. The reason is quite simple. Sometimes when fillings are removed, more mercury escapes into the body than if the fillings were left untouched. More importantly, when amalgam is removed, there is almost always a temporary increase in the amount of mercury in the blood. For pregnant women this significantly increases the risk of higher mercury levels in their fetuses."

Witness D: A field researcher of dentists and dental practice
"We must be careful of dentists who are eager to pull these fillings. Some dentists who have lost their business to other dentists could use this situation to increase their business. They could be so anxious to get this new business that they get sloppy and careless. In doing so they could increase the mercury escaping into the human body. Unfortunately, there are dentists who would engage in such practices."

Witness E: A practicing dentist
"I don't have the slightest hesitancy in using amalgam. It has proved safe and effective for nearly 150 years. I want nothing but the best for my patients, including children and pregnant women. Indeed, if I had any doubts about its safety or concerns about its possible harm to my patients, I would refuse to use it."

Witness F: A member of the Food and Drug Administration
"It was not until the mid-1980s that the potential of mercury poisoning from amalgam became an issue. Now that we have available more accurate instruments to measure the escape of mercury vapor in one's mouth, we can collect better data. We don't know what the amounts of vapor mean for human health. We don't know enough at this time to stop the use of amalgam or to issue a warning to the general public. Certainly we welcome all the scientific studies we can get. We need data on the actual effects of the amalgam in the mouths of humans."

Witness G: A representative of several teams of Swedish investigators studying amalgam fillings
"We have studied over 2,500 people and found no correlation between the number of amalgam fillings and symptoms of mental disorders or other illness."

Witness H: A representative of a medical team
"We have screened the health of a large number of dentists who have frequent direct exposure to amalgam fillings. Our results reveal that dentists are healthier than the general population. When there is an increase in the mercury levels, we have found the cause was the diet of these dentists rather than their exposure to amalgam."

Witness I: A chemist for a dental research team
"The whole issue is not worth discussing. It is my estimation that within 5 to 10 years there will be many substitutes for amalgam. These will work better and many will be less expensive than silver amalgam fillings. Some current substitutes are gold, porcelain inlays and crowns, and porcelain resin composites."

You have heard these witnesses and now have a chance to study their comments. You must make a decision as to what should be done.

In a preliminary meeting, the panel considers some of the alternatives available to it. The panel could decide to:

1. Ban all future use of amalgam within the community, starting immediately and for all time.
2. Delay all action on the use of amalgam until extensive research confirms that it is indeed harmful. This research is likely to take many years.
3. Encourage members of the public who have amalgam fillings to have their blood checked for mercury.
4. Ban all future use of amalgam within the community, starting immediately and until their is convincing evidence it is safe.
5. Forbid doctors from replacing amalgam fillings in patients to reduce the likelihood of mercury poisoning.
6. Forbid the use of amalgam fillings for all pregnant women, starting immediately and for all time.
7. Encourage dentists to reduce their use of amalgam on a voluntary basis but place no limitations on its use.
8. Require the city to spend thousands of dollars on a public-awareness campaign of the health risks of amalgam. This would be similar to the warning on cigarette packages. Dentists would be allowed to use amalgam unless the patient specifically asks for something else.
9. Place a special tax of $100 for each amalgam filling a dentist puts in the mouth of a patient.

There are a number of other options that could be selected. The panel is directed to make a decision after carefully examining the data. As individuals or as a group, complete the Predecision Task Sheets. Then make a final decision. The panel is free to make any decision it believes is appropriate. Panel members may stick to the options above or make any other single or set of recommendations it wants to make.

PREDECISION TASK SHEET 1

1. The major problem to be solved in this situation is

2. The two strongest reasons this is the major problem are
a.

b.

3. From this source, what are at least three possible effects of mercury poisoning on human health?
a.

b.

c.

4. From these testimonies, what evidence is there that amalgam is dangerous to human health?

5. From these testimonies, what evidence is there that amalgam is not dangerous to human health?

6. What is the name of the person(s) or organization(s) who gave the *most* believable testimony?

7. What two reasons make this person or organization's statements so believable?
   a.

   b.

8. What is the name of the person(s) or organization(s) who gave the *least* believable testimony?

9. What two reasons make this person or organization's statements not believable?
   a.

   b.

10. What would you consider to be the very worst part about this entire situation?

11. In this situation and the search for a solution, what should be the roles of science and scientists?

12. In this situation and the search for a solution, what should be the roles of technology?

After you have completed your individual answers, share your responses with the others in your group. Add to, revise, or correct your responses as needed. Make sure everyone in your group comprehends this information before going on to the next page.

PREDECISION TASK SHEET 2

In the space below, write the three options you consider the most appropriate at this time. Decide the possible consequences of these options for the public and dentists. Then do the same thing for the three options you consider the least appropriate. Remember, you are not restricted to the options in the story. If you are aware of better options or worse options, list them.

Most Appropriate Options at This Moment	Possible Consequences of This Option for the Public	Possible Consequences of This Option for Dentists
1.	a.	a.
	b.	b.
2.	a.	a.
	b.	b.
3.	a.	a.
	b.	b.

1. The three reasons these are the *most* appropriate policies to select and follow in this situation are
   a.

   b.

   c.

Least Appropriate Options at This Moment	Possible Consequences of This Option for the Public	Possible Consequences of This Option for Dentists
1.	a.	a.
	b.	b.
2.	a.	a.
	b.	b.
3.	a.	a.
	b.	b.

1. Three reasons these are the *least* appropriate policies to select and follow in this situation are
   a.

   b.

   c.

After you have completed your individual answers, share your responses with the others in your group. Add to, revise, or correct your responses as needed. Make sure everyone in your group comprehends this information before going on to the next page.

INDIVIDUAL DECISION SHEET

Before you work with other members of the panel, make a set of decisions on your own. You need to construct responses to the questions and statements below. This task will help you consider the situation, problem, and possible solutions in a systematic way.

1. After considering the situation and the options available, the very best decision to make at this time is to

2. The three most important reasons this is the *best* policy to follow are

   a.

   b.

   c.

3. Two major benefits of this policy for the public are

   a.

   b.

4. Two major benefits of this policy for dentists are

   a.

   b.

5. The group that will be most affected by this decision is
__________________________.
Two important reasons they will be affected most are
a.

b.

6. The information from the testimonies that was the *most important* in influencing the decision was

7. The very *worst* thing that could be done is

because

When everyone in your group has completed answers to the above, move on to developing a group decision. Use the Group Decision Sheet to record your panel's decisions.

GROUP DECISION SHEET

At this point your entire group will construct a single consensus response to this situation. Everyone must help make the decision and must agree on each decision. Arrive at your group responses by consensus, not by majority rule.

1. After considering the situation and the options available, the very *best* decision for us to make at this time is to

2. The three most important reasons this is the *best* policy to follow are

a.

b.

c.

3. Two major benefits of this policy for the public are

a.

b.

4. Two major benefits of this policy for dentists are

a.

b.

5. The group that will be most affected by this decision is ______________________________.
Two important reasons they will be affected *most* are
a.

b.

6. The information from the testimonies that was the *most important* in influencing the decision was

7. The very *worst* thing that should be done is

because

REVIEW AND REFLECTION QUESTIONS

Suggested follow-up questions to focus and guide inquiry and learning.

a. What is *mercury?*

b. From other sources, what dangers does mercury pose for human health?

c. What are the symptoms of mercury poisoning?

d. Amalgam contains mercury and mercury has been known for years to cause health problems in humans. Why would the Food and Drug Administration approve its use in 1976?

e. From the data in the story, what information was collected by scientists?

f. From the data in the story, what information was provided by the group you found to be most biased?

g. From the data in the story, what information was the most scientific?

h. How many amalgam fillings are in your mouth? The mouths of members of your family?

i. To what extent did the number of amalgam fillings you have in your mouth influence your analysis of these data about mercury poisoning?

j. To what extent did the number of amalgam fillings you have in your mouth influence your final decision as to what should be done?

k. Suppose you had a friend who suffered effects of mercury poisoning from amalgam fillings. How might your decision in this situation have been different?

l. What specific types of research would convince you that amalgam fillings do or do not pose a real danger to human health?

m. Suppose your dentist told you that amalgam posed no danger to human health. How would his or her statement affect your decision to use amalgam?

n. Suppose one dentist in your community said that amalgam posed a real danger to health, and another said there was absolutely no danger. Which one would you believe?

o. Why would you believe one dentist over the other?

p. In situations such as that in this story, what should be the role of scientists to resolve such issues?

q. If you were a scientist and were asked to conduct research on amalgam, what specific questions would you ask to guide your research?

r. This story is about how people make decisions regarding what research is worth paying attention to and how people will use research to support their position. In this story, what evidence is there that people used research findings to support their existing views?

s. In this story, how seriously did you consider information that was different from the view you already had about amalgam fillings?

t. If many dentists are already biased in favor of using amalgam, to what extent can you believe their interpretation of research findings about amalgam?

u. Given the information in this story, if you have an amalgam filling in your mouth and you can afford to have it replaced, will you have it replaced?

v. Given the information in this story, if you need to have a filling, would you want to consider alternatives to amalgam fillings? How would you go about researching such alternatives?

EPISODE **21**

A HAREY SITUATION

Dr. Manuel Ramirez was late for work. He had stopped by the hardware store to get a box of bullets. The bullets were to go with the surprise birthday rifle for his son, Juan.

As Dr. Ramirez entered the back door of his office, he heard the barking of his anxious, impatient, unhappy patients. Dr. Ramirez was a respected veterinarian in this small farming community. His patients were the animals people brought into his office or that he encountered on visits to homes and farms. Dr. Ramirez smiled and said "Good morning" to his staff, patients, and Juan.

Juan spends as much time as he can at his father's office. He is his father's assistant. He enjoys caring for the animals who have to spend the night in small, confining cages. He is learning to detect visible signs of improvement in sick animals. He also knows many of the danger signals of when their health gets worse. Juan spends long hours comforting animals that are especially sick. He does not believe any animal should be hurt or killed on purpose.

A HAREY SITUATION

This Saturday morning, Juan did not mind cleaning the cages and washing the food containers. It was his birthday. He knew his father's office was only open until noon. This morning, he noticed his father had hurriedly hid a large box.

Finally, the last patient had been treated. Only Juan and his father were in the office. Both felt the excitement of Juan's birthday.

Dr. Ramirez carefully handed Juan a long, neatly wrapped package. Juan shouted with delight. His eyes glistened with awe as he held the brand new .22 rifle. He had been using an old rifle the past several months during target practice. He instantly guessed the contents of the small package his father then handed him.

Dr. Ramirez and Juan rushed to their jeep. The biggest surprise was yet to come. They stopped by Sam Miller's house. He and his son mounted their rifles in the rack. In a few minutes, Juan realized they were all going hunting—not just target practicing, but real, live rabbit and hare hunting. He tingled with excitement.

Juan knew a great deal about rabbits and hares, including the fact that they are not the same animal. **Rabbits** are native to Europe, the Americas, Asia, and Africa. Rabbits are smaller than hares, tend to be gregarious, and are quite adept at gnawing. They have long ears, short tails, and usually gray or brown fur. They are prolific breeders—mature females can give birth to three or more litters a year, of two to eight babies each. Their gestation period is 28 days. Their babies are born blind, furless, and helpless. The babies are cared for in the nest or burrow. Rabbits live in grass nests or burrows, either of which may be new or abandoned by other animals. Rabbits can also transit tularemia (rabbit fever) to people.

Hares are native to central and southern Europe, Africa, North America, and the Middle East. Hares are generally solitary animals. They have ears usually longer than their heads, well-developed hind legs, and a keen sense of hearing. Their body length is about 40 to 70 centimeters (16 to 28 inches) at maturity, not including their short tail. In northern areas, they usually have white fur in the winter and grayish-brown fur in the summer. Elsewhere they usually wear gray-brown fur all year round. Jackrabbits are hares, not rabbits. Hares are born fully furred with open eyes. They begin to hop a few minutes after birth.

Rabbits and hares are mammals of the order *Lagomorpha* and the family *Leporidae.* They are both vegetarians, although some individual rabbits and hares have been known to eat meat.

Juan smiled as he thought about what he knew about rabbits and hares. He was also glad that, unlike most people he had met, he knew rabbits and hares were not the same animal.

A HAREY SITUATION

After the jeep bumped over rutted dirt roads, the hunters arrived at a familiar destination. In the spring, Juan had often enjoyed walking over this now barren, unproductive farmland. He and his dad had once found some arrowheads buried in the edge of the wooded areas. This area was now overrun with hares and rabbits.

Hundreds of rabbits and hares had eaten the garden vegetables and grass seedlings. The farmers could not protect their crops and the environment was nearly ruined. In some regions, coyotes and wolves were natural predators of hares and rabbits and helped keep a balanced environment among species. But there were no coyotes or wolves in this area. Efforts to get rid of the hares and rabbits, even with poison, had been unsuccessful. Unless most of these pests were eliminated, the fertile farmland could not be used.

The four hunters quickly reviewed the safety rules of hunting. Then they walked deeper into the surrounding woods. Juan began to realize what was happening. He had been so excited about his birthday, and had felt so grown up to be hunting with his dad, that he had not really thought about what was now expected of him.

The foursome silently approached the section known to contain rabbit and hare burrows. Juan was told he would have the first shot.

Suddenly Juan spied an adult hare before him. He raised and leveled his rifle, set the sighting, and prepared to pull the trigger. The hare turned to face Juan. For a brief second their eyes met. Thoughts and glimpses of the animals at his dad's office flashed through Juan's mind.

He was paralyzed as he watched the hare scamper into the nearby thicket. Tears came to his eyes. He knew he had disappointed his dad. Crying would only make it worse.

He blamed the quickness of the hare for his failure to shoot. He spent the rest of the afternoon listening to his father and Sam Miller talk over the reasons for shooting rabbits and hares. Deep inside Juan knew he would not be able to shoot his new rifle today. He sensed, for the first time, a problem he could not easily solve. Today he may have gotten by without shooting, but what about the next time and the next time?

Juan has to make a decision.

Imagine that you are Juan. You must make this decision. If you go hunting again, your father and other people will expect you to shoot the rabbits and hares that are destroying the environment in the area. Before you make your decision as to whether you will shoot them, review the information you have learned. Rethink what you believe about killing animals. To guide you in this task, complete the Predecision Task Sheet and the Decision Sheet.

PREDECISION TASK SHEET

To help you review relevant information before making your decision, answer the questions below.

1. According to the story, what is Juan's attitude toward animals?

2. According to the story, what is Juan's attitude toward killing animals?

3. Of the facts given about *rabbits,* what are five you hope you never forget?
   a.
   b.
   c.
   d.
   e.

4. Of the facts given about *hares,* what are five you hope you never forget?
   a.
   b.
   c.
   d.
   e.

5. In this story, what specific problems are the people having with rabbits and hares?

6. To what extent might these environmental problems be caused by people rather than by rabbits and hares?

7. In this situation, what are three possible decisions Juan could make?

   a.

   b.

   c.

8. While this story was not specifically about science and the work of scientists, a number of details could be directly linked to science. What are three areas in the story that could be linked to science?

   a.

   b.

   c.

9. While this story was not specifically about technology, a number of details could be directly linked to technology. What are three areas in the story that could be linked to technology?

   a.

   b.

   c.

After you have completed your individual answers, share your responses with the others in your group. Add to, revise, or correct your responses as needed. Make sure everyone in your group comprehends this information before going on to the next page.

INDIVIDUAL DECISION SHEET

1. Suppose you were Juan and this story was about you. Suppose your dad asked you to explain why you did not shoot the hare. What would you have told him?

2. What are your two strongest reasons for your decision?
   a.

   b.

3. If you were Juan, the best thing for you to do the next time your dad wants to take you hunting is

4. What are your reasons for this decision?

5. Suppose someone said, "There are times when it is necessary to kill animals to protect the environment." How would you react?

6. Pretend you are a classmate of Juan's and that he has asked for your advice about what to do the next time he is invited to go hunting. What are three important ideas you would share with Juan?
   a.

   b.

   c.

7. The best advice you could give Juan is

GROUP DECISION SHEET

Imagine Juan asks you, a group of his friends, to help him in his decision-making process. He wants your group to make one set of answers to the questions below. Make your decision by consensus, not by majority rule.

1. Imagine Juan's father asked Juan to explain why he did not shoot the hare. What should Juan have told him?

2. What should be Juan's two strongest reasons for this decision?
   a.

   b.

3. If you were Juan, the *best* thing for you to do the next time someone wants to take you hunting is

4. What are your reasons for this decision?

5. Suppose someone said, "There are times when it is necessary to kill animals to protect the environment." How would you react?

6. The very best advice you can give to Juan regarding hunting and shooting hares and rabbits is

7. In this situation, the very *worst* thing for Juan to do is

REVIEW AND REFLECTION QUESTIONS

Suggested follow-up questions to focus and guide inquiry and learning.

a. In the story, what was Dr. Ramirez's area of medicine?

b. What are four important jobs a veterinarian does?

c. What were three jobs Juan did at his father's office?

d. If you could summarize the main job of a veterinarian, what would it be?

e. What are two things Juan and his dad have in common?

f. What does the phrase *species control* mean?

g. What does it mean to *hunt?*

h. In what situations might a person go hunting without going out to kill something?

i. What stopped Juan from shooting the hare?

j. If you had been Juan's father, what would you have said to Juan after he failed to shoot?

k. In what ways does a rabbit differ from a hare?

l. How might shooting at a paper target be different from shooting a live animal?

m. How might species control techniques lead to extinction?

n. If you have ever been on a hunt, at the time you saw your target, what were your feelings?

o. In what ways might species control be a form of conservation? Ecological protection?

p. What makes hunting a sport?

q. In what ways might it be good for people to kill animals that are pests?

r. In the story, if the rabbit population continued to grow, what would happen to the farmlands?

s. At what point would a species or pest be "under control"?

t. From what you know, what are some other situations where species control techniques are needed?

u. In what ways might one claim that a gun is a piece of technology? A product of science?

v. Besides killing, what are two other ways people could get rid of pests?

w. If you had been along with Juan on this hunting trip, would you have shot the hare? Several hares?

x. The title of this story is "A Harey Situation." What part of the story would most fit the title?

y. Suppose you were a farmer and your crops had been destroyed by hares and rabbits. Suppose you had been responsible for helping get rid of the coyotes that lived in the area. If someone wanted to reintroduce coyotes to cut down on the rabbit and hare populations, would you still be against the coyotes?

z. What are three important ideas this story tells you about the ways animals and the natural environment can affect one another?

EPISODE **22**

WHOSE FAULT?

Last year Benson Construction Company won the bid for construction of the Tate Nuclear Power Plant. The negotiations had been long and tiring, following the U.S. Atomic Energy Commission's final approval. The services of Brooks Engineering Company have been retained by the power company. Within one month, an affiliate of Benson Construction, Pico Landscaping, will begin preparing the 19,000-acre tract of land. The actual plant facility will fill 400 of the 19,000 acres.

The main nuclear plant will house the pressurized water reactor in the containment building. Lake White will be the major supplier of water. Two new dams will be constructed. One dam is for the large, main reservoir. The second dam will provide water for the small bay reservoir, where the water will be cooled.

Everyone thought that the power company's site selection study had been thorough. The environmental, engineering, and scientific reports included lots of geology, ecology, and archaeology data.

Trenching had been done to determine the various geological composites.

A few well-informed citizens in the neighboring community of Troy have been vocal in their attempts to stop the plant's construction. They looked everywhere for help in explaining alternative energy sources. Experts in geothermal power, solar energy, and wind power spoke at night meetings.

Proposals for producing geothermal power for the Troy area had no supporters. The only geological areas now known to have natural hot water streams for turning turbine generators are in the western United States. The solar energy experts prepared many reports. While hating to admit it, they reluctantly agreed that the cost of electricity from a solar power plant was too high at the present time. In desperation they provided studies and various methods of home conversion to solar energy systems. This, too, was impractical for many homes.

While wind-power electrical generators were enthusiastically explained, they were impractical for this low-elevation location. Besides, vibrations and noise emitted from the turning blades convinced few that this choice was feasible.

The opponents to nuclear power did not convince the officials of the Tate Power Company to stop the nuclear power plant.

The past several months showed extremely slow progress in all aspects of construction. The land where the waste-processing building will be is just now being leveled.

This morning the first shift of workers arrived at 6:30 a.m. Some of them, including Richmond Reid, drove 40 miles to work. They entered the temporary gate, received security clearance, and scattered to various locations for the day's work.

Once inside the gate, Richmond and his fellow workers traveled about 3 miles, past the main containment building to the future site of the waste-processing plant. He was soon in a hole of bedrock shouting measurements of layers of clay, shale, sandstone, and conglomerate up to his partner. These measurements were critical because the waste-processing plant would temporarily store the radioactive wastes in underground tanks.

At that moment, Richmond realized there was a break in the layers of sandstone. The drop difference was 7 feet. The layers remained constant on both sides of the horizontal crack. Richmond could not believe that previous investigations had not revealed evidence of this fault. If the fault continued, an earthquake might occur. If the nuclear wastes were stored underground on the fault line, there would be severe leakage of wastes along all levels of the sedimentary layers. To make certain his instruments were correct, he measured the area three more times. He used his partner's equipment in case his instruments were giving inaccurate data. Each time the information matched his original measurements.

Richmond knows how far behind schedule construction is. There are deadlines to meet; the officials ordered a speed up in production. Richmond must make a decision.

He knows he must continue to take accurate measurements and pictures of the fault. However, he is not sure what he should do with the facts he has gathered. He considers his options.

1. He could notify the construction company manager, but he remembers that anyone who delays construction will be fired on the spot.
2. He could notify the power company engineer, but the engineer has seen some of the data already and may have chosen to ignore it.
3. He could bypass the company bosses and notify the local townspeople, but this means he would be fired and may never be hired by other companies.
4. He could notify an ecologist he knows, but any news of the fault could be linked directly back to him.
5. He could remain quiet about the fault and hope a miracle prevents an earthquake on the fault line.
6. He could tell his fellow workers about the fault, but they are even more powerless than he is.
7. He could (add your own option)

8. He could (add your own option)

9. He could (add your own option)

Richmond is not sure what he should do. *Suppose you were Richmond Reid in this situation.* What would you do?

Before you make a final decision, answer the questions in Predecision Task Sheets 1 and 2. Then it is up to you to make the best decision you can make. Record your final decision on the Decision Sheet.

PREDECISION TASK SHEET 1

Take time to search through reference materials to find answers to the following questions. You may work on these answers by yourself or with members of a small group.

1. What is the definition of a *fault?*

2. In terms of length and depth in kilometers, how large can a fault be? How small can a fault be?

3. According to scientists, what causes faults to occur?

4. If an earthquake were to occur, what could happen to the land along a fault line?

5. What is an earthquake?

6. What geological events might bring about an earthquake?

7. What does a nuclear power plant look like?

8. What specific damage could a major earthquake cause to a nuclear power plant?

9. To what extent can scientists accurately predict when an earthquake will occur?

10. What are acceptable reasons for building anything along or very near a fault line?

11. What specific parts of this story and the work of Richmond Reid can be linked directly to science?

12. What specific parts of this story and the work of Richmond Reid can be linked directly to technology?

After you have completed your individual answers, share your responses with others in your group. Add to, revise, or correct your responses as needed. Make sure everyone in your group comprehends this information before going on to the next page.

PREDECISION TASK SHEET 2

What happens at the power plant will affect many people in many ways. The chart lists some of those people and has space for you to add other people or groups. As an individual or member of a group, write some of the possible consequences of continued construction.

Affected People	Possible Positive Consequences of Continued Construction	Possible Negative Consequences of Continued Construction
1. Townspeople	a.	a.
	b.	b.
2. Richmond Reid	a.	a.
	b.	b.
3. Richmond's fellow workers	a.	a.
	b.	b.
4. Tate Nuclear Power Plant officials	a.	a.
	b.	b.
5. Benson Construction Company officials	a.	a.
	b.	b.
6. Visitors to the power plant	a.	a.
	b.	b.
7. Future workers at the power plant	a.	a.
	b.	b.
8.	a.	a.
	b.	b.

After you have completed your individual answers, share your responses with the others in your group. Add to, revise, or correct your responses as needed. Make sure everyone in your group comprehends this information before going on to the next page.

INDIVIDUAL DECISION SHEET

1. Suppose you are Richmond Reid. After considering all the information, you have decided the very *best* thing to do is

2. The single most important reason for this decision is

3. What are at least three *positive* consequences of this decision?
   a.
   b.
   c.

3. What are at least three *negative* consequences of this decision?
   a.
   b.
   c.

4. In this situation, the very *worst* thing Richmond Reid can do is

5. What are at least two ways Reid's decision may affect the scientists working with this project?
   a.
   b.

6. What are at least two ways Reid's decision may affect the technology used with this project?
   a.
   b.

7. Suppose the data about the fault are revealed and the power plant and construction company managers and employees choose to complete the project. Suppose 5 years later, an earthquake partially destroys the plant, endangering the townspeople. What specific persons or groups should be held responsible?

GROUP DECISION SHEET

Suppose you are a group of Richmond Reid's closest friends. He shared this information with you and asked your group to arrive at one set of responses to the items below. Reach your decisions using consensus, not majority rule.

1. After considering all the information, the very *best* thing for Richmond Reid to do is

2. The single most important reason for this decision is

3. What are the three most important *positive* consequences of this decision likely to be?
   a.
   b.
   c.

4. What are the three most important *negative* consequences of this decision likely to be?
   a.
   b.
   c.

5. What are the two most important ways this decision will affect scientists working on this project?
   a.
   b.

6. What are the two most important ways this decision will affect technology used on this project?
   a.
   b.

7. Suppose the data about the fault are revealed and the power plant and construction company managers and employees choose to complete the project. Suppose 5 years later, an earthquake partially destroys the plant, endangering the townspeople. What specific persons or groups should be held responsible?

REVIEW AND REFLECTION QUESTIONS

Suggested follow-up questions to focus and guide inquiry and learning.

a. In what ways is a geologist a scientist?

b. What kinds of things do geologists as scientists study?

c. Suppose there were geologists living in your community. What kinds of information could they give you that would be relevant to you?

d. Suppose a geologist said it was not his or her job to predict earthquakes. What would be your reaction to this claim?

e. Suppose a geologist predicted an earthquake and it didn't occur. What would be your opinion toward this geologist? Toward all geologists?

f. What are at least three reasons a company would hire a geologist?

g. Suppose a company hired geologists and these scientists provided information different from what the company expected. What obligations does the company have to pay attention to this scientific information?

h. Suppose in the story you were the owner of the power plant. If the geological information indicated your plant was near a fault line, what would your decide about building and opening the plant?

i. You read about scientists who have important information that people will likely interpret very differently depending upon their position or need. In these cases you can see how people and societies influence what will be done with scientific information. What are three other areas in which people have very different views as to what the scientific data mean or what should be done with the data?

j. If you were the owner of this plant, what other scientists might you call to help offset the information about the danger of the fault?

k. What is the closest fault line to the community in which you live?

l. One of the most massive earthquakes in recorded history occurred in the United States. Along what major river did this earthquake take place?

m. In the United States, nuclear power plants have been built within a hundred miles of active or potentially active fault lines. Knowing what we know about the power of earthquakes, why would Americans build nuclear power plants in such locations?

n. Some nuclear power plants were built along the coastline of major oceans or along the banks of major rivers. If radiation were to escape from these plants, what might happen to water life affected by this radiation?

o. What roles could scientists play in the location and building of a nuclear power plant?

p. Why would a person want to become a geologist?

q. What are at least three places you would likely find a geologist working?

r. In what ways have geologists contributed directly or indirectly to the way you live?

s. What would be the most exciting thing about being a geologist? The worst thing?

t. What does a geographer do?

u. What are at least two major differences between what geologists and geographers do?

v. What would keep you from being a geologist?

w. In this story there were major concerns that the scientific data might be ignored. Why would people ignore important scientific information? What would be the important reasons you would ignore scientific data?

x. There is a large quantity of scientific data on the effects of smoking on the health and lives of humans. Yet many people ignore these findings and continue to smoke. Why would people ignore these scientific data?

FLIGHTY DECISIONS

Josh Caputo, director of Skyport International Airport (SIA), must make a decision. For the last five years, the number of people departing and arriving from international flights at SIA has slowly but steadily declined. Inflation, the recent recession, and increased fares have partly been to blame for the decrease in passenger load. As if that weren't enough, national passenger loads have also decreased, but at less rapid a pace. Several of the less important airlines using the terminal facilities are in danger of going out of business. Some of these airlines have applied for government loans and subsidies. Meanwhile, Josh has seriously considered reducing staff at the facility and closing segments of the terminal to reduce costs.

In the midst of these concerns, Josh received word from Washington that his terminal was one of a few airports eligible to accept flights of the new foreign-built supersonic aerotransport (SAT). The SAT may be the end of Josh's troubles. It is capable of bringing in, on a single flight, 500 passengers—over twice the number of passengers of most conventional jet aircraft. Not only would such a new plane bring in and take out larger quantities of people, but increased volume may lower overall passenger fares—a fact likely to bring even more people to the airport. Early trial flights from Europe to South America have supported these possibilities. Even more encouraging is the fact that the plane flies over 850 miles per hour. This speed is nearly 300 mph faster than all subsonic passenger airplanes.

Josh is aware that it would take nine to twelve months to ready his terminal for such flights. He would need to purchase equipment and special machinery. He would need to hire and train a crew to maintain and service the aircraft. A segment of the terminal would have to be renovated to meet the demands of the larger numbers of people coming in on and departing from the SATs. He would need to study and follow federal guidelines regarding these operations. Gaining acceptance of regularly scheduled SAT passenger flights will require a great deal of work and money.

At this point, the proposed benefits far exceed the expected initial expense. Should the SAT give the boost needed, Josh will not have to lay off workers or close down more segments of the terminal. In addition, public interest in cheaper and faster SAT flights is expected to generate renewed interest in air travel in general. The danger of the smaller airline companies going out of business will be lessened. More jobs will be saved. In fact, many new jobs will be created.

Josh also realizes the criticisms from various sectors of the population must be reviewed and considered in his decision to accept or reject the SAT flights. Among the most important of these criticisms are the following:

- The increased pollution generated by the SAT will speed up the destruction of the Earth's ozone layer.
- While an SAT can carry many more passengers on a single flight, there is no guarantee more people will want to fly in one.
- The SAT burns more fuel than the conventional jet liner; hence in an era when fuel conservation is needed, we have built another guzzler.
- The SAT would cut flight time by approximately 35 percent.
- The possibility is great that large groups of people needed to fill one of these planes will not want to go to the same place at the

same time; hence one of the major advantages of such a large plane will not be realized.

- The SAT has not been fully flight tested to eliminate all possible bugs.
- The sonic boom generated by the SAT will add to the noise pollution level and may cause property damage.
- There is no guarantee how long the SAT flight service will continue should it prove to be an economic failure.
- The quality of life for those persons living near the airport will diminish further.
- The cost per fare on early SAT flights will be the same as regular flights on smaller airplanes, but with increased use, the rates are expected to be cut.

Josh is aware some of the SAT critics have not been consistent in their use of their arguments. One of the strongest anti-SAT groups has argued on protect-the-environment grounds, yet its members have not used such arguments against stopping flights of supersonic bombers into and out of a major Strategic Air Command (SAC) base just a few miles from Skyport International. Josh's comments regarding the inconsistency of their use of air and noise pollution, energy use, and ozone destruction as criteria for condemning the SAT have gone unanswered.

Today when Josh arrived at his office, he received a call from his district's congressional representative. He was informed there was a possibility—almost a certainty—that should he accept the SAT flights the federal government would pump money into Skyport International to support the cost of SAT flights during the first year of operation. Money would be allocated to renovate the terminal, hire and train service personnel, and ready the airport for the flights. Because of the lateness of the session and the possibility of the available money being used to support another project, the representative cannot wait long for Josh's decision. Josh must inform him of his decision within three hours. Josh agrees to make up his mind and call the representative.

As Josh contemplates his decision, he hears two news items over the radio. First, one foreign nation is completely satisfied that the latest trial flight of the SAT went well. "All kinks have been worked out. The SAT is a marvel," said the foreign spokesperson. Second, a group of scientists is convinced that the frequency of cancer will increase significantly as the ozone layer is destroyed.

Needing to make his decision, Josh calls a special meeting of the SIA Advisory Board. He informs them of the problems and the situation. He tells them of his need to make the decision. At this time he is told that over the past six months

- 710 workers of SIA or the airline companies working out of SIA have been laid off.
- SIA had $6.9 million less income than for the same period one year ago.
- Rising electricity rates will continue to increase the operational expenses of the airport.
- 77,000 fewer passengers departed from the airport than departed during the same period one year previously.

As Josh receives this news, the phone rings. The representative reminds him he has about two hours left.

You are a member of the SIA Advisory Board. You are to help Josh make his decision. If you recommend a decision and it fails, you will be partly to blame and will probably not get recommended for reappointment to this position in the future.

Josh agrees to follow the decision of your group, the SIA Advisory Board. To ensure that you consider the major points, Josh asks you to do the following things:

- Examine the major issues involved in accepting the SAT flights into SIA. To do this, use the Predecision Task Sheet.
- After each individual has reviewed the issues and data, make an individual choice as to what ought to be done. Following the statement indicating your personal choice, write down the major reasons for your choice on the Individual Decision Sheet.
- Once all members have made their individual choices, reach a group decision through consensus, and write it on the Group Decision Sheet. Agree as to the most important reason that will justify your decision.

PREDECISION TASK SHEET 1

To prepare for the decision you must make, take time to answer the questions below. You may work with others to arrive at your answers.

1. According to the story, what factual information should be used *to support* SAT flights into SIA?
   a.

   b.

   c.

   d.

2. According to the story, what factual information should be used *against* SAT flights into SIA?
   a.

   b.

   c.

   d.

3. By accepting the SAT flights, what three *benefits* are likely to come to the community?
   a.

   b.

   c.

4. By accepting the SAT flights, what three *areas of harm* could come to the community or the environment?
   a.

   b.

   c.

5. By accepting the SAT flights, what are three likely benefits *for* the airport?
   a.

   b.

   c.

6. By accepting the SAT flights. what are three likely benefits *lost* for the airport?
   a.

   b.

   c.

7. What are three specific details about the SAT and SAT flights that are directly linked to science and scientists?
   a.

   b.

   c.

8. What are three specific details about the SAT and SAT flights that are directly linked to technology?
   a.

   b.

   c.

After you have completed your individual answers, share your responses with others in your group. Add to, revise, or correct your responses as needed. Make sure everyone in your group comprehends this information before going on to the next page.

PREDECISION TASK SHEET 2

Before you make your personal choice, take time to think of four choices you could make. Write at least one benefit and one disadvantage of each option.

Possible Decisions	Likely Benefits of This Decision	Likely Disadvantages or Negative Effects of This Decision
1.	**a.**	**a.**
	b.	**b.**
2.	**a.**	**a.**
	b.	**b.**
3.	**a.**	**a.**
	b.	**b.**
4.	**a.**	**a.**
	b.	**b.**

INDIVIDUAL DECISION SHEET

For this decision, form responses to the statements below on your own.

1. After reviewing the information, I recommend the SIA Advisory Board make the following decision:

2. The three *best* reasons I can give for supporting this decision as the choice to make are
   a.

   b.

   c.

3. The three *best* possible results of this decision are
   a.

   b.

   c.

4. The three *worst* possible results of this decision are
   a.

   b.

   c.

5. The *worst* decision I could make is

GROUP DECISION SHEET

For this task, work as members of the SIA Advisory Board. Work together to form one set of decisions. These decisions are to be reached by consensus, not by majority rule. Make sure everyone in the group agrees with each decision before it is written down.

1. After reviewing the information available and discussing our individual decisions, as the advisory board to SIA, we recommend the following course of action be taken:

2. The three *best* reasons we can give as a justification for our group decision are
   - **a.**
   - **b.**
   - **c.**

3. The three *best* possible results of this decision are
   - **a.**
   - **b.**
   - **c.**

4. The three *worst* possible results of this decision are
   - **a.**
   - **b.**
   - **c.**

5. The very *worst* thing we could do is

REVIEW AND REFLECTION QUESTIONS

Suggested follow-up questions to focus and guide inquiry and learning.

a. From the information given, how did you become a member of the SIA Advisory Board?

b. Suppose you refused to accept the SAT flights, and the airport continued to lose flights and passengers. What would your attitude then be about accepting such flights?

c. Suppose the government ordered Josh to accept SAT flights. If you were Josh, how would you respond to this action?

d. Why would environmentalist groups that oppose the SAT not also oppose the building of supersonic bombers?

e. Is it a good or bad policy for the federal government to support programs that may destroy the environment?

f. If you were one of the laid-off airport employees, what decision would you hope the SIA Advisory Board makes?

g. Suppose the SIA Advisory Board could not agree on a decision and the representative called back demanding a decision. If you were Josh, what would be your decision?

h. What is the major theme of this story?

i. What is the major problem Josh and the airport officials had to resolve?

j. In what ways was science involved in the problems the airport officials had to solve?

k. In what ways was science involved in helping resolve the problems at the airport?

l. In what ways did the SAT represent science? In what ways did it represent technology?

m. What is a least one way science and technology differ?

n. If you were against the airport accepting SAT flights, how might you use science to stop such flights?

o. If you supported the airport accepting SAT flights, how might you have used science to help your cause?

p. Besides noise and air pollution, what other ways might a crowded airport affect the environment?

q. What are at least three benefits of improved size and speed of passenger planes?

r. If you lived near SIA, what would your reaction be to the possibility of SATs flying over your house?

s. How might the location of a person's house or apartment or the type of job a person has affect their decision about SATs landing at a local airport?

t. If SATs flew over your house or apartment two or three times a day, what would your feelings be toward those who allowed them to land?

u. Suppose that low-flying airplanes over your house or school were military rather than passenger planes. What words would best describe your feelings toward the pilots of those planes?

EPISODE **24**

HEAR YE! HEAR YE!

The annual spring party and dance is the biggest event of the year. Because poor bands have played in the past, most students stopped going to the dance that follows the party. This year, however, the dance promises to be great. Gator Bait, the best band in the area, will play. Everyone is talking about going to the dance.

Imagine you are a student at Hontoon Intermediate School. Imagine you are student chairperson of the dance committee. With the help of several teachers and students, you have coordinated the entire event. You contacted the band. Three weeks ago you were congratulated by your principal for your hard work.

It is two weeks before the party and dance. Every detail has been checked and double checked. During second period you are called to the principal's office. He had just received a phone call.

"A group of students from the university has been studying noise levels," said the principal. "At a dance last week, Gator Bait's music reached a noise level of 116 decibels for long periods of time. As you know, this level is harmful to the human hearing system. When the students informed the band of the noise level, they laughed. In fact, they played even louder. I am very close to stopping Gator Bait from playing at our dance. The music they play is too loud. But it is too late to get another good band."

The principal suggests that you, as the representative of the student body, decide what should be done. You are to think about the situation, report to him after lunch, and tell him your decision at that time. As you try to decide, you remember the following:

- After years of poor attendance at the dance, this year most students are eagerly awaiting the dance.
- The talk among students for weeks is that they find it hard to believe you actually got Gator Bait to play at the school dance.
- The annual spring party and dance is the big event of the year, especially for those students who will leave the school in June.
- Many students worked hard to get the money to hold the party and to hire the band.
- Your science teacher had just reviewed the information about noise, decibel levels, and particular types of hearing problems caused by extremely loud sounds, including sounds made by rock bands and radios played at high volumes.
- Some teachers and parents oppose the dance because they don't like the type of music played by the band.
- Music played too loud is known to cause permanent and extensive damage to the ears.
- You have been congratulated by nearly everyone for your outstanding work and leadership. You are pleased with everything you have done, especially in getting the band.
- You have been active in pro-environment programs and activities, especially those concerning air and noise pollution. Until now, you had not made a connection between the band's music and noise pollution.
- Turning down the band's music might lose you some friends.
- It is too late to get another good band.
- The principal can still cancel the band if you don't.

Use the Predecision Task Sheets to consider important information related to your decision. Complete both Predecision Task Sheets and share your responses with one or more members of your class. Then you will be ready to record your response to the principle on the Decision Sheet.

PREDECISION TASK SHEET 1

A special system is used to indicate the range of loudness a person can hear. The *decibel* is the unit of measure between the lowest and the highest range of sound. Study the information below. Place an X on the space to the left of the sounds you hear on a typical day.

Sound Level in Decibels		
140	___	Painful, extremely damaging sound
130	___	Sonic boom, jackhammer (3 feet away), air-raid siren (100 feet away)
120	___	Very loud thunder, some rock bands (within 40 feet)
110	___	Airplane jet engine (within 100 feet)
100	___	Subway train (20 feet away), heavy truck (25 feet away)
90	___	City buses (10 feet away)
80	___	Orchestral music, pneumatic drill, vacuum cleaner
70	___	Noisy office, ordinary traffic, noisy classroom
60	___	Normal conversation, window air conditioner
50	___	Quiet automobile
40	___	Quiet office
30	___	Quiet conversation
20	___	Whisper (5 feet away)
10	___	Rustle of leaves
0	___	Beginning of hearing

Take some time to reflect upon the Xs you made on the chart. What do these Xs reveal about the amount of noise in your life?

PREDECISION TASK SHEET 2

1. What is a *decibel?*

2. On the decibel scale, what is the real meaning of a difference in 10 decibels between two levels of sound?

3. Suppose you were a member of an environmental action group. How would you define *noise pollution?*

4. At what point does *music* become *noise?*

5. At what point does *noise* become *music?*

6. What makes certain levels of sound so dangerous to humans?

INDIVIDUAL DECISION SHEET

1. The two *most important* reasons I might keep this band are

a.

b.

2. The two *most important* reasons I should find another band are

a.

b.

3. After considering the information on decibel levels and the reasons I wrote above, the best decision I could make is to

4. The two *most important* things that will happen because of my decision are

a.

b.

5. I would explain my decision to the students in this way:

6. I would explain my decision to the principal in this way:

7. I would explain my decision to the members of my school's environmental action group in this way:

8. The *very worst* thing I could do in this situation is

9. The two reasons this would be the worst thing to do are

a.

b.

GROUP DECISION SHEET

Imagine that the principal selects a group of students to help you make this decision. The principal wants a consensus decision. As a group, respond to the items below.

1. The two *most important* reasons we might keep this band are
   a.
   b.
2. The two *most important* reasons we should find another band are
   a.
   b.
3. After considering the information on decibel levels and the reasons we wrote above, the best decision we could make is
4. The two *most important* things that will happen because of our decision are
   a.
   b.
5. We would explain our decision to the students in this way:
6. We would explain our decision to the principal in this way:
7. We would explain our decision to the members of our school's environmental action group in this way:
8. The *very worst* thing we could do in this situation is
9. The two *most important* reasons this would be the worst thing to do are
   a.
   b.

REVIEW AND REFLECTION QUESTIONS

Suggested follow-up questions to focus and guide inquiry and learning.

a. According to the principal, what damage is anticipated from the loud music?

b. Before making your decision, which facts did you consider?

c. What are at least three possible consequences of your decision?

d. Before your made your decision, what alternatives did you consider?

e. Suppose you were also chairperson of your school's environmental action group. How might this position influence your decision in this situation?

f. At what point does music become noise?

g. At what point would music, noise, and pollution be identical?

h. When faced with making a decision to protect the things you enjoy or the things known to protect the environment, which is more important?

i. When you were told you had to make this decision, what were your feelings?

j. What is your definition of *pollution?*

k. At what point could music played by a rock band be considered noise pollution?

l. Suppose someone played a portable radio too loudly. At what point would you consider its sound noise pollution?

m. To what extent is it possible to restore damage to a person's hearing mechanisms?

n. In what ways would the decibel scale be a tool of science? A product of science? A product of technology?

o. Of what value to you are such tools as the decibel scale?

p. It is scientifically confirmed that sounds in the 110+ decibel range are very dangerous to your health and hearing. To what extent will this information affect your decision to listen to music played this loudly?

q. If you know loud music is very likely to cause severe damage to your hearing and you still listened to it, what might this say about the value you place on scientific findings?

r. What kinds of information would people need to have to persuade them not to listen to music played at levels above 90 decibels?

s. Why do some people like to have music played at levels above 110 decibels?

t. Suppose at the age of 20 or 30 you found your hearing fading because of your listening to very loud music now. Who would you blame for this problem?

u. Using the results of scientific studies and new technology, electronic music companies are making amplifiers that make sounds that consistently reach or surpass the 130 to 140 decibel range. To what extent is such equipment an example of a good use of scientific information?

v. In our society many people believe we should prevent people from doing things that could be harmful to them. If someone prevented you from listening to music that could cause severe damage to your hearing, what would your feelings be toward that person?

w. Why would a band play music above 100 decibels?

x. What right do members of a band have to play music at a level they know is likely to severely affect the hearing of many of the people who listen to it?

y. If you were a member of Gator Bait and were directed to play your songs below the 100-decibel level, what would your reactions be to the directive?

Exploration Decision Episodes

Exploration decision episodes provide students with a limited set of subject matter to consider in written or graphic form. A set of questions helps students consider the information they are to examine, clarify, and use. The questions also help monitor student subject matter comprehension, how students relate the subject matter to the focus of the inquiry, and what decisions and judgments students have made about the information and issue presented.

THE CASE OF THE MISSING STATUE

The famous detective, Mali Cruz, hung up the phone. She turned to her assistant, Dr. Dixon, and said, "Well, another mystery faces us today."

"And what's that, Mali?" asked Dr. Dixon.

"The rock musician's gold statue has been stolen. This statue, worth thousands of dollars, was given for having a number-one album and single at the same time. Few singers ever achieve this level of popularity. We must find out who stole the musician's statue." Cruz *stated the problem.*

Dixon's face showed concern. "Why must we do it? Can't the police find out?"

Cruz calmly reminded Dixon that the musician was a close family friend who had recently had financial problems. "Besides," she added, "the police are baffled and haven't been able to crack the case. They have been working on it for six months and have no clues. I just wish the musician had called me earlier.

"The statue is a priceless and irreplaceable gift to the musician's family from a British king. Anything that is stolen should be located and returned," Cruz reasoned. "The thief should be brought to justice. These are reasons enough. Our task is to find the statue and the thief. Come, Dixon, it is time we began our investigation."

After hearing the purposes of the investigation, Dixon asked Cruz whether she had any idea or clue as to who the thief might be.

Slowly, Cruz *began making a number of hypotheses* (plausible inferences or guesses) about who may have stolen the statue.

"Well, from what I have learned so far, it seems to be an inside job," Cruz said. "It wasn't a break-in. I presume the thief to be employed by the musician—maybe even a servant or a friend. It could have been a visitor to the musician's mansion. The thief had to have detailed knowledge about the house and the statue. Let's get our things together, Dixon. The musician has invited us to be his guests until we can solve the mystery."

"I'll pack the magnifying glass and the fingerprint kit. Is there anything else we need?"

"Yes," replied Cruz as she *thought through the needed materials and equipment.* "We'll need my diamond-studded watch and matching ring. And bring your revolver. The musician said he was held at gun point from behind during the robbery. We may need to protect ourselves."

As Dixon and Cruz rode to the musician's mansion, Dixon questioned Cruz. "What's your plan? I can understand the fingerprint kit and the gun. Your personal jewelry has confused me."

Cruz *described the procedure or the way she planned to go about solving the mystery.* "Remember, Dixon, the police have thoroughly questioned all the servants. They have come to no conclusions about who the thief is. I shall play the thief's game. Sometimes, to catch a criminal, one must think like a criminal. If the thief has the idea that the police have given up, he or she may consider himself in the clear. Feeling safe, the thief may slip up and give himself or herself away. I expect to give that person a chance. I will set a trap for our man or woman!"

Dixon and Cruz were greeted by the musician. Immediately the two investigators *began making observations and collecting relevant data.* They heard the details concerning the holdup and possible suspects. The three visited the room where the statue had been.

The maid was cleaning in the room where the robbery had taken place. She noticed an empty wine glass on the bookshelf, and was annoyed to find spices in powder and leaf fragment form on the carpet. She apologized for not noticing these before today.

After several hours, the musician called all the servants together. He announced they were all free to go about their business because the police were looking elsewhere for the robber.

Dixon and Cruz were introduced as house guests. As Cruz and Dixon greeted the servants, one servant asked the musician to use his master key to unlock the room of another servant who had entered the hospital the previous day.

"John wants his book and personal journal. You know no one else can get into our rooms," the servant remarked.

Throughout the afternoon, Cruz rubbed her fingers over the antique diamond watch and ring. Later that evening, after dinner and a long conversation with the musician, Cruz and Dixon went to their adjoining rooms. Once safely in their rooms, Dixon called Cruz.

"What on earth were you doing, Mali? All afternoon and evening you drew attention to yourself, playing with that jewelry of yours. Everyone kept staring at your diamond watch. It was very impolite. Very impolite, indeed! Then you gave it to the musician to put in the very safe the statue was stored when it was stolen."

Cruz calmly *reported her observations.* "Dixon, rather than watching me you should have been watching the eyes of the servants. Certainly all eyes were on me and my jewelry. Only one pair of eyes stayed fixed on me, and they brightened when the musician reported that no one in the mansion was considered a suspect. I questioned the musician about this person—his name is Mr. Rexford. He is the chef and, recently, a frequent loser at gambling. Lately, Mr. Rexford has asked for advances on his wages to pay his gambling losses. Remember that the maid found small traces of spices, sage, cumin, and oregano near where the musician kept his statue. Also only one person drank more than one glass of red wine—Rexford. Now we just have to wait."

"Wait for what? It's time to go to bed."

"Come on, Dr. Dixon. Bring your revolver. As you know, my diamond watch and ring are in the cabinet safe where the musician kept his statue. We have to be ready to spring the trap."

They remained in their positions in the dark room near the safe until 4:00 a.m. Suddenly, they heard a noise! Someone was in the room. After a minute had passed, Cruz turned on the lights.

Dixon was surprised. "Why, it's Rexford, the chef! He has the key to the cabinet safe. I've got him covered, Cruz. Call the police."

As the musician entered the room still wearing a bathrobe, Cruz said, "Unfortunately, Dixon, you are pointing the gun at the wrong person." She turned to the musician. "Isn't he?"

Dixon was confused. "What do you mean? If Rexford isn't the man we want, why was he going for the safe in the middle of the night?"

Rexford told them he found the key to the safe on the kitchen table. He took it with the intentions of stealing the watch and ring to help pay off his gambling debts.

Dixon had made his conclusion based on the information as he had interpreted it. "There, Cruz, you are wrong! Rexford is the one we want. He tried to steal your jewelry, and he stole the statue."

Cruz *carefully made her conclusion.* "You are partly right, Dixon. Rexford did try to steal my jewelry. However," Cruz said as she started to *give evidence and an explanation to support the full conclusion,* "it was the musician that made it easy for him by leaving the key out. The musician knew Rexford needed money because the musician himself had gambling debts. Until yesterday afternoon Rexford never told the musician why he needed advances in money. But the musician suspected gambling debts and knew they had been for small amounts. There was no reason for a servant to steal an expensive statue for small debts. Besides, it would be nearly impossible for a servant to sell the statue.

"It was the musician who had very heavy debts and had to take out several loans. The musician sold his own statue and reported it stolen to get the insurance for it. He hoped to pin the blame on Rexford here by encouraging the theft of my jewelry. That would throw full suspicion on Rexford and put himself in the clear.

"Only the musician has a master key. He used it to get the wine glass from Rexford's closed room and the spices from the locked cabinet in the kitchen. The maid cleans and vacuums the house every weekday. There is no way she would have missed the glass on the bookcase and the spice powder and leaf fragments on the floor all these weeks. They were planted this morning as evidence against Rexford. The musician's trick has backfired. *This problem is solved.* Call the police, Dixon. We have our thief!"

The story above illustrates one set of steps one might follow in investigating problems. Some scientists might use similar steps in their work. These steps are also used by others involved in trying to solve problems and doing investigations. However, these are not the only set of steps a scientist may use to solve problems.

REVIEW AND DECISION TASKS

Take time now to answer the questions below. Be ready to give reasons for your answers.

1. In this story, Cruz went through a number of major steps to solve the mystery. In the order given, what were these major steps?
   a.

   b.

   c.

   d.

   e.

   f.

   g.

2. According to the story, what is another word for *hypothesis?*

3. If Cruz had not carefully observed the servants as she was introduced, what might have been the consequences for the chef?

4. What are two important reasons it is important to carefully record observations made during an investigation?
   a.

   b.

5. What are three ways the work of a detective is similar to the work of scientists?

   a.

   b.

   c.

6. What specific information in this story would help you become a better scientist?

7. If you were the musician, what words would you use to describe the way Cruz conducted her investigation?

REVIEW AND REFLECTION QUESTIONS

Suggested follow-up questions to focus and guide inquiry and learning.

a. What is the major theme of this story?

b. What does this story have to do with science?

c. In what ways would Mali Cruz be a scientist?

d. How is a conclusion connected to a hypothesis?

e. What did Dixon fail to do that caused him to accuse the chef?

f. Why is it a good idea to have someone check your procedures and observations before you make a final conclusion?

g. What makes something a problem?

h. What was the major problem Cruz and Dixon had to resolve?

i. What is an *investigation?*

j. What is a *procedure?*

k. What information did you find in this story that you could use in handling problems in your own life?

l. To what extent was Dixon a scientist?

m. To what extent was Dixon a close observer of what was occurring around him?

n. What was it that enabled Cruz to solve the case and Dixon to accuse the wrong person?

o. Suppose someone said this was a science story as well as a detective story. Would you agree or disagree with this statement?

p. If the findings that result from your scientific investigation are different from your hypothesis, what should you do?

q. In what situation in your life might you use these particular steps of investigation?

r. Of these steps of investigation, which is the most important?

s. What are the names of other types of scientists who might make use of this method of finding solutions to problems?

t. When doctors try to diagnose an illness, which specific steps in this method of investigation would they follow?

u. How is conducting an investigation using these steps of investigation similar to the steps you would follow to resolve a problem such as noise pollution?

v. In this story, what was the major problem? Why was this a problem?

w. What other methods of investigation could Cruz have used?

x. What other methods of investigation could a scientist use?

y. What does this story tell you about how people in your community might resolve certain problems?

EPISODE **26**

A BIRD STORY

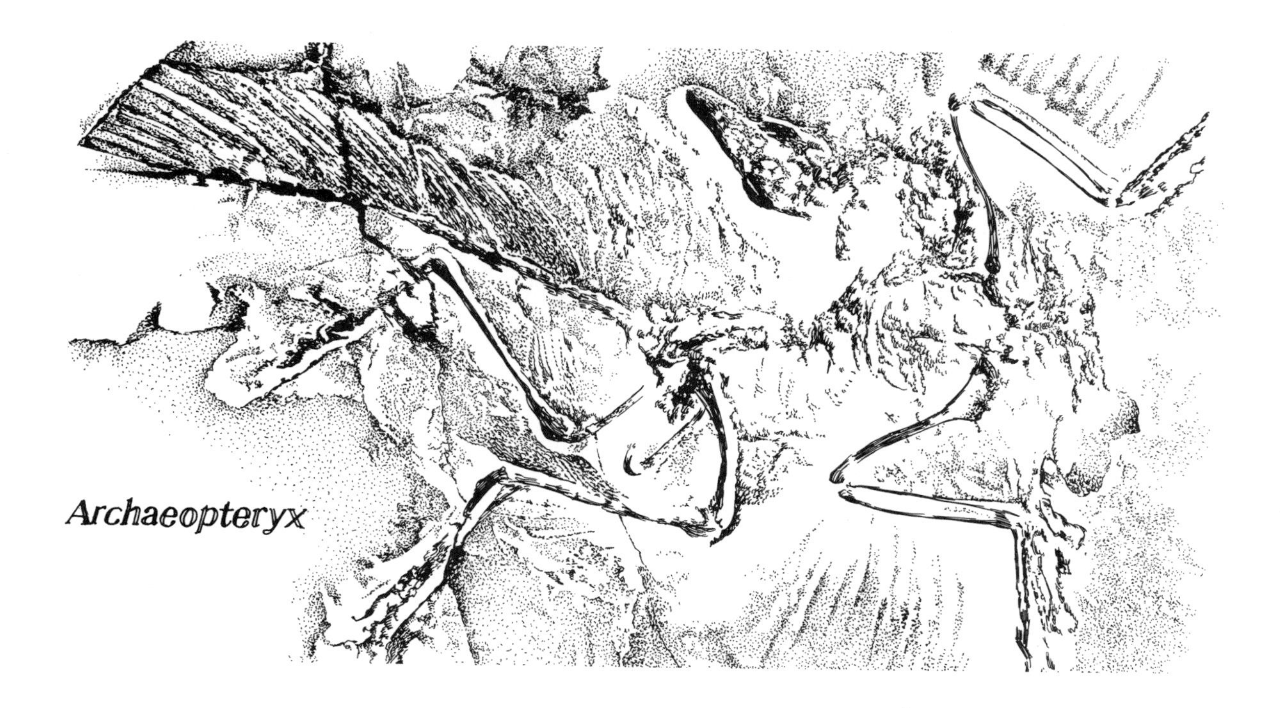

Press Release

After 100 years paleontologists are still not sure whether the first birds were reptiles or mammals. Some evidence suggests the earliest birds were cold-blooded, like reptiles, and capable of only short flights before overheating.

The first preserved record of an early bird was uncovered in the limestone quarries of Bavaria, Germany, in 1861. This bird, called *Archaeopteryx* or "ancient wings," started a controversy among scientists that has still not ended. Since 1861, five skeletons and a single feather of this bird have been discovered.

This creature was about the size of a pigeon. It was, however, far more primitive than today's birds. When the Archaeopteryx was first unearthed, most scientists considered it to be a small dinosaur, complete with pointy teeth, three fingers on each claw, and a long, bony tail. The earliest birds had a skeleton and structure that were not aerodynamic. They were related to the dinosaurs in that both these birds and the dinosaurs were reptiles. However when other fossil remains of the creature were found, there was clear evidence that it had feathers. The questions then centered on whether this creature actually could fly and, if so, how.

A new analysis of the muscle mass and skeletal features of a 145-million-year-old bird originally lead to a very controversial conclusion.

Dr. John Ruben, a zoologist at Oregon State University, claimed during an October 1990 conference that the earliest birds were not warm-blooded. He claimed these birds were physiologically reptiles, but still birds. They had feathers. They flew. But they were cold-blooded. Present-day birds are warm-blooded.

Per square inch, the muscles of cold-blooded animals such as lizards and snakes are twice as powerful as the muscles of warm-blooded animals such as dogs and modern birds—even humans. According to Ruben, the earliest bird had too small a skeleton to be able to carry enough warm-blooded muscle to enable it to fly. But if it had the muscle strength of a cold-blooded animal, it could have gotten airborne.

At the same conference, two other scientists, Paul Sereno and Cheng-gang Rao, announced the discovery of a new fossilized bird that lived 135 million years ago—10 million years after the Archaeopteryx. This fossil, unearthed in 1987 in Liaoning province in northeastern China, was revealed after three years of study by the two paleontologists. This creature, they claim, was clearly a flying, perching bird.

Today most paleontologists believe today's birds are so similar to dinosaurs that it is a mistake to say the dinosaurs died out 65 million years ago. To them today's birds are direct descendants of earlier dinosaurs.

1. What do scientists mean by the term *cold-blooded?*

2. What is the definition of a *fossil?*

3. What is the definition of a *bird?*

4. Given the information in this story, in what way is an Archaeopteryx a bird?

5. Why is it important to know whether the first bird was a dinosaur or mammal?

6. From the information in the story, how would you answer this question: Was the first bird a reptile?

7. What is the major theme of this "bird story"?

8. What evidence would you need to confirm that the first bird was a dinosaur or mammal?

9. How could paleontologists claim today's birds are direct descendants of the dinosaurs?

10. In what ways is it good that scientists continue to study fossils?

11. Suppose scientists did determine that modern-day birds are descendants of dinosaurs. How would this discovery affect the people in this society and you?

REVIEW AND REFLECTION QUESTIONS

Suggested follow-up questions to focus and guide inquiry and learning.

a. What is a *reptile?* What is a *bird?*

b. In what ways are present-day birds and reptiles different from one another? Identical to one another?

c. According to the story, in what country was the first early bird fossil found?

d. What does the name *Archaeopteryx* mean?

e. What does a paleontologist do?

f. The information about these fossils was announced during a conference of paleontologists. What are two important purposes of such conferences?

g. What kinds of topics would one have to study to become a paleontologist?

h. In what ways is it good for a society to have paleontologists conduct research on fossils?

i. What other interpretations could be made of these data?

j. How can scientists tell that a particular fossil is 145 million years old?

k. Suppose the scientists had five fossils skeletons of the same creature and did not have a fossil feather. How might their interpretations of this creature have been different from what it now is?

l. Some people argue that "the facts speak for themselves." Is it ever possible for fossils to "speak for themselves?"

m. What fossils have you seen? What did these fossils look like?

n. What should scientists do with the fossils they find?

o. In terms of years, how old does something have to be to be called a fossil?

p. Suppose a person finds a fossil in a national park. To whom does that fossil belong?

q. At what point in time should a society tell scientists to stop searching for fossils?

r. Where could you find fossils in or near your own community?

s. If you could find one fossil, what would that fossil be?

t. Why would you want to become a paleontologist?

u. Why would you not want to become a paleontologist?

v. How might scientists use technology to help them solve the mystery of the first birds?

w. What does this article reveal about the ways scientists view the same data?

ICEBERG!

Scientists from New Zealand and the United States have completed their observations of an iceberg. They had tracked a single iceberg for 3 years. The iceberg, 96 miles long by 22 miles wide, broke away from the Antarctic coastline's Ross Ice Shelf.

During the 3-year, 1,250-mile trip, the floating ice block was tracked using satellites and transmissions received from the radio transmitter dropped on the berg. The iceberg's voyage was a western route along Antarctica's coast. The study revealed the existence of four separate ocean currents in the region. This study verified the nature and extent of these ocean currents, which were thought to exist according to the widely accepted ideas on ocean currents.

One important reason for the study was to check out ideas concerning how effectively oceans have been and are absorbing carbon dioxide. Polar currents are believed to hold critical information about the Earth's climate. The absorption of carbon dioxide has been linked to theories concerning global warming.

Some scientists believe the oceans are now close to carbon dioxide saturation. If that happens the oceans can no longer absorb carbon dioxide from the atmosphere and convert it into much-needed oxygen for animal life on the surface.

One way to find out if this idea about oceanic saturation is correct is to track ocean currents and find out how ocean waters flow. In the Antarctic region cold currents flow downward under warmer ocean currents, parallel to the ocean surface. If cold currents continue to absorb carbon dioxide from the atmosphere and take it below the surface, saturation hasn't occurred. Should they stop absorbing carbon dioxide or stop flowing downward, saturation has occurred, and our planet might be in a dangerous state of contamination.

Scientists on the project said their interest in ocean currents in Arctic regions of Earth is not new. In July 1893, the Norwegian explorer Fridtjof Nansen sailed in the Arctic Ocean on the *Fram.* This small ship was specially built to cross the Arctic Ocean without being crushed by the tremendous pressure of the ice pack and to withstand serious damage should it brush against an iceberg.

Nansen made the journey after analyzing the results of previous studies of ocean currents in the area. He concluded that a current flowed from a point north of Asia across the North Pole before turning south from the Svalbard islands to Greenland. His hazardous voyage on the *Fram* included many changes in direction. Many of the *Fram*'s crew believed they were trapped in eddies and backwaters and would never survive the ordeal. Eventually, the ship, locked in ice, was carried hundreds of miles from the North Pole to Franz Joseph Land by February 1896. There the currents carried the *Fram* south, where it was finally freed as the ice that trapped the ship melted. Nansen's ideas about the Arctic currents were verified.

The study of this iceberg may be the last of its kind. Satellite technology is expected to replace the need to monitor iceberg drifts from ships. Infrared and other types of photographs taken from satellites are targeted to be the workhorses for the study of oceanic currents. Other satellites with other remote-sensing devices are expected to provide continuous information about the ocean.

1. What is an ocean current?

2. In this study, scientists tracked an iceberg 96 miles long by 22 miles wide. If the center of this iceberg were in your classroom, what locations in your state would mark its outer boundaries?

3. What are three reasons these scientists were studying this iceberg?
   a.

   b.

   c.

4. According to this report, what is one major connection between icebergs, ocean currents, and global warming?

5. How might your personal behavior in your local community affect icebergs and ocean currents?

6. What could you do to prevent contamination of the atmosphere?

7. To what extent should scientists continue to study icebergs and ocean currents?

8. What specific pieces of technology might these iceberg tracers have used in their research?

9. Suppose a citizen of your community criticized the government for spending money on studying one iceberg for 3 years. What would you say in response to his or her criticism?

REVIEW AND REFLECTION QUESTIONS

Suggested follow-up questions to focus and guide inquiry and learning.

a. What is an *iceberg?*

b. What are at least two reasons scientists would need to study an iceberg for three years to get information about ocean currents?

c. In what ways is the submerged part of an iceberg similar to the hull of a sailboat in water?

d. If the submerged part of an iceberg and a hull of a sailboat are similar, what might this information mean for the route the iceberg will likely take in the ocean?

e. In what ways is the exposed part of an iceberg similar to the above-water parts of a sailboat?

f. If the exposed part of an iceberg and the above-water parts of a sailboat are very different, what might this information mean for predicting the influence of the wind on the route of an iceberg?

g. What are at least two important findings of the voyage of the *Fram?*

h. What are at least two important reasons for studying the two polar regions of Earth?

i. What type of person would spend three years in the Arctic or Antarctic studying icebergs and ocean currents?

j. To what extent would you like to spend three years of your life in such a location?

k. If you could spend three years of your life intensely studying something, what would you want to study?

l. These scientists spent three years tracking one iceberg. What does this time period reveal about the work of some scientists?

m. The report reveals that scientists sometimes spend many years just collecting data. If these scientists relied more on satellite data-collection technology, would they need to spend more or less time collecting identical data?

n. Looking at a map of the world, what are the routes of at least three major ocean currents?

o. What major ocean current is closest to the community in which you live?

p. What is *global warming?*

q. What are the most likely causes of global warming?

r. What are the most likely effects of global warming on life on Earth?

s. What are at least two reasons scientists disagree as to whether global warming is or is not occurring?

t. What does this disagreement among scientists about global warming reveal about scientists? About science?

u. What scientific evidence would you need to see to convince you global warming is or is not occurring?

v. In this story, what specific types of technology were mentioned?

w. In what ways did the scientists use technology to help them in this study?

x. A space satellite will be used in the future to collect data on icebergs and ocean currents. Is such a satellite technology or an example of technology?

y. In which ways could the ship *Fram* be an example of technology?

ATHLETES' FEATS

You and Marcia had always had a special friendship. As children you spent hours playing and talking together. As you grew older, you enjoyed playing tennis and attending track meets together. Several years ago, Marcia and her family moved from your neighborhood. You miss jogging with her in the late afternoon. Now you only see her at track meets and races. You had a strange feeling the first time Marcia beat you in a long-distance race. You never thought of her as a rival. However, you began seriously studying techniques to improve your running.

You remember that when you first began jogging, you went through combined intervals of walking, jogging, and resting. Even now, such combinations take the drudgery out of training. At first, you really had to concentrate to jog slowly. You worked to an established pace in your training and continued that pace. It was essential to form this habit to be able to pace yourself during a race.

To strengthen your leg muscles, you repeated your stretches, kicks, and bends many times. Living in a hilly neighborhood was helpful, and you forced yourself to run close to home at least once a week. Other days, you ran at the track to work on timing and endurance. You were consistent in your daily running. You were aware of the importance of breathing exercises, and you concentrated on pacing yourself as you took oxygen into your lungs.

Your serious attitude and your diligent training have been effective. You finished ahead of Marcia in a 10-kilometer race. You talked to her following the race, but the cool friendliness made you sad. Yet both of you realized you had to continue competing against yourself and each other.

The established pattern continued. In another town, Marcia finished before you in a race. In yet another race, you outran Marcia. These races were practices for the 42.2-kilometer (26 miles 385 yards) marathon, the biggest event of the year. You know Marcia will be competing. You have to win. You cannot help but think that now is the time to stop these back-and-forth wins.

You read that most runners fast one to three days a month, and for several months you practice various methods of fasting. Water and juice are the only things consumed on fasting days. You find that if you fast for 24 hours prior to a race instead of only 3 to 6 hours, you run faster and with more energy.

All liquids consumed during a race or any type of exercise should be kept at 5 degrees Celsius (41 degrees Fahrenheit). You learn that packaged sports drinks are better to drink than water during a race. They resupply your body with needed minerals. In fact, sweating during exercise robs your body of vital fluids and electrolytes. Without these, the body cannot maintain the needed levels of efficiency and energy.

You read that sports drinks are scientifically formulated blends of carbohydrates and electrolytes designed to get fluids and nutrients into the body fast. They contain glucose, sucrose, and sodium, which help speed fluids through the blood stream and provide energy to working muscles. They contain no caffeine and no fat. Like most runners and athletes, you value the way sports drinks help you when you exercise, practice, and compete.

You also learn about the 6-day carbohydrate-loading plan. The first 3 of the 6 days prior to a race, you eat high-protein foods. This drains the energy reserves stored in carbohydrates. The next 3 days just before the race, you eat only foods high in carbohydrates. These foods quickly restore your energy level, allowing you to run faster.

Some runners rely on vitamin-mineral loading. This second

method seems only to give some endurance and better post-race recovery.

You wonder what Marcia is doing to get ready for the race. You know her well enough to know she is as determined as you are to win.

The day of the race finally arrives. You do not see Marcia as the starting gun is fired. As the crowd of runners thins out, you see Marcia ahead of you. You remind yourself to pace your speed. You have to catch up to Marcia. Your legs begin to feel like heavy weights. As Marcia reaches out for the sports drink handed to her from one of the designated persons, you pass her.

The halfway mark finds you still ahead of Marcia. You seem to be breathing too hard. At this point, you shouldn't have this problem. You concentrate on being more aware of your pacing. Perhaps, without realizing it, you have been running faster. As you attempt to even out your pace, Marcia passes you. She smiles as she gives you a look of, "I understand. I too feel the pain. I too must pace myself. I too want to win."

After 39 kilometers, you are limping. Your left knee feels like it wouldn't bend again. You see Marcia stumble. She seems to be having trouble with her ankle. Often runners are able to detect the most painful areas in their competitors, probably because they have felt the same pains and know how they themselves had reacted. This most often happens when competing against someone the runner knows relatively well.

As you get close to the finish line, you realize Marcia will win. Although she has slowed down, there is no way you can run any faster. As she turns to look at you, you exchange painful glances. Hers is one of pain and happiness. Yours is one of pain and sadness. Your knee will hardly bend at all. But you must cross that finish line, even if it is behind Marcia.

One final spurt of determination moves you closer to Marcia. She turns around again. As she does, she reaches out her hand to you.

You grasp it. Together you cross the finish line.

1. What is the main theme of this story?

2. It was once remarked that there is "a science to running and finishing a marathon." What did the person mean by this statement?

3. How might a drink for athletes be scientifically formulated?

4. What are the most common ingredients in sports drinks?

5. To prepare for and to acquire sufficient nourishment for a marathon, what are two plans, besides drinking liquids, some runners follow?

6. If you were Marcia in this story, as you crossed the finish line, what would be your feelings about your accomplishment?

7. If you had run this race and Marcia had helped you finish number one alongside her, how would you feel toward Marcia?

REVIEW AND REFLECTION QUESTIONS

Suggested follow-up questions to focus and guide inquiry and learning.

a. Why would scientists become involved in formulating a drink for athletes?

b. What kinds of research might the scientists who formulated sports drinks have done to develop them?

c. Would sports drinks be an invention, discovery, or creation of scientists?

d. What is an *athlete?*

e. Besides drinks, what other ways have scientists helped athletes perform better?

f. In what ways have scientists reduced the level of athletic performance?

g. For many years athletes were directed to swallow lots of salt tablets before and while they were exercising. To what extent is the consumption of high levels of salt recommended for today's athletes?

h. What is an "athletes' feat"?

i. How would scientists determine which of the plans described the story was the best one for marathon runners to follow?

j. Besides the area of food and beverage, to what other areas could scientists contribute to help athletes perform at higher levels of efficiency?

k. Suppose you were an athlete. How might you go about studying the diet that would be most helpful for you?

l. Suppose you were an athlete. How might you go about studying the practice routines that would enable you to perform at your highest levels?

m. Suppose you were an athlete. What would keep you from studying your diet and practice routine so you could become a better athlete?

n. How have scientists helped ensure that athletes do not use illegal substances to enable them to perform better?

o. Is it good for scientists to work to improve the performances of athletes?

p. To what extent is it good for scientists to develop ways of detecting when athletes use illegal substances?

q. If you were an athlete, how much time would you spend reading about the scientific studies that could improve your performance?

r. If you were an athlete and scientists found you were doing things that would limit your performance, how much attention would you pay to these scientific findings?

s. How might a sports drink be considered an example of technology?

t. What are three ways athletes use technology to improve their performance?

u. To what extent should society support the efforts of scientists to improve the performance of its athletes?

v. In what five ways might athletes use technology to improve their performance?

w. Outside of athletics, what are three pieces of technology you use to improve your performance in everyday situations?

x. Sports equipment, such as the modern tennis racket, is sometimes considered a technological tool rather than merely a piece of equipment. If this is the case, how can we determine whether increasing performance is the result of the person or the technology?

y. If you had been Marcia in this race, would you have reached back and grabbed the hand of a close friend behind you?

z. If you had been in second place in this race and Marcia had extended her hand to you, what would you have done?

BREAKTHROUGHS

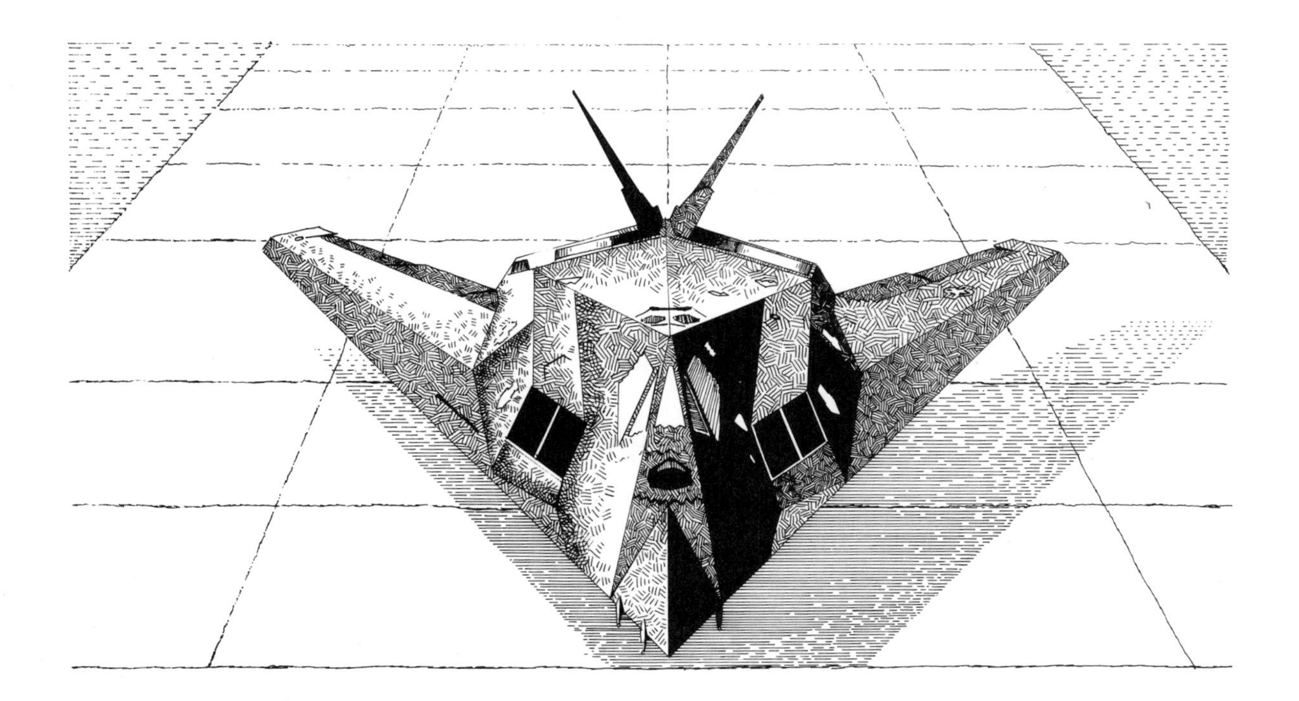

The room was very quiet. The speaker began his report. "The Pentagon is ready to reveal one of its most well-kept secrets. These concern the design and construction breakthroughs leading to the radar-evading Stealth or B-2 bomber."

He continued, "The future of the Stealth bomber is uncertain. As civilians, we have no control over the decision regarding its future. However there is much to be learned and used from the technology that went into its design and construction. You are here to consider the knowledge we have gained from this project, and to consider whether and how you will make use of this knowledge. The technology should have wide applications in future air and space projects. It may have many uses in civilian and commercial projects.

"The B-2 bomber is sturdy enough to carry 20 tons of bombs and missiles and light enough to fly 10,000 miles with one midair refueling. Many of its metallic parts, including engine covers and other heat-resistant parts, were formed from titanium alloys. People in the aerospace industry have long favored titanium. It is lightweight and can withstand intensely high heat. However, it has several disadvantages. Titanium is difficult to work with because it is brittle. It snaps when bent and is difficult to stretch and form.

"To overcome these difficulties, scientists at LTV Aircraft adapted a technique called *superplastic forming* to make titanium panels as large as 4 by 12 feet. This is nearly triple the size of the titanium panel produced for earlier aircraft.

"Superplastic forming molds brittle metal at high temperatures over long periods of time, making it pliable and easier to shape and bend. A U.S. Air Force project in the 1970s lead to this process. Superplastic forming helped produce smaller titanium parts for other aircraft, such as F-15 and F-18 fighters, as well as the Stealth.

"To make the Stealth work as expected, LTV Aircraft had to reduce the number of bolts, rivets, and other metals that stiffened and fastened the airplane's outer shell and engine shields. This was a high priority because enemy radar can detect planes easier as the number of fasteners increases. Eliminate these outer parts, and the plane is less likely to be picked up by radar.

"To further improve the plan, LTV used a process called *diffusion bonding* to enable builders to combine two large titanium parts produced on a press to form a larger, more complex part. With this process and press, many complex parts can be made and combined into single, very large, very complex parts. To make the titanium panels, a 1,000-ton press was used to produce sheets that weigh as much as 150 pounds with a quarter-inch thickness.

"Inside the presses, the titanium sheet is heated to 1,650 degrees Fahrenheit. At that temperature, the rock-hard titanium softens to a puttylike consistency. Then argon gas is forced into the press. Flowing at pressure rates of 300 pounds per square inch, the gas forces the titanium onto a mold the shape of the wing part, or any shape needed. Then the titanium is allowed to cool.

"There are disadvantages of superplastic technology. The costs for retooling a production line are very high. Unless there are large numbers of sales or someone is willing to pay extremely high prices for a few items, manufacturers will never make a profit. There may be some uses for it in building large airplanes such as the Boeing 747 or the new aerobus that expects to carry 900 passengers. At present, however, it seems this technology will have few uses outside these major aero-industry projects."

1. What is the defination of *superplastic forming?*

2. What advantages does superplastic forming have over other methods of molding titanium?

3. What is the definition of *diffusion bonding?*

4. What is the definition of a *breakthrough?*

5. What information in this report is directly linked to the notion of a breakthrough?

6. What is the definition of *technology?*

7. In the superplastic forming process, what specific roles would technologists play?

8. In the superplastic forming process, what specific roles would scientists play?

9. In what specific ways would the roles of scientists and technologists be different in the superplastic forming process?

10. What three ideas about how a society can affect the use of technology are illustrated in this story?
    a.

    b.

    c.

REVIEW AND REFLECTION QUESTIONS

Suggested follow-up questions to focus and guide inquiry and learning.

a. According to this article, what is one major reason why the Stealth bomber is likely to evade being detected on radar?

b. What is an *alloy?*

c. Why would one want to use titanium to build aircraft?

d. In your own words, what is involved in the process of making parts of the wings of a Stealth bomber?

e. How difficult would it be for this company to make titanium wings for other aircraft?

f. In what ways does science differ from technology?

g. What are at least three ways science and technology are related?

h. Who determines whether something is science or technology?

i. Who determines whether or not members of a society will be allowed to make use of the latest scientific discoveries?

j. Who *should* determine whether or not members of a society will be allowed to make use of the latest scientific discoveries?

k. What individuals should determine how members of a society will make use of the latest scientific discoveries?

l. Currently, who determines whether or not members of this society will be allowed to make use of the latest technological discoveries?

m. What individuals *should* determine whether or not members of this society will be allowed to make use of the latest technological discoveries?

n. Who should determine how members of this society will be allowed to make use of the latest technological discoveries?

o. What are at least five ways that uses of technology can directly affect you?

p. What are at least five ways individuals like yourself can directly affect the uses of technological devices?

q. What are at least four benefits to society of improving technology?

r. What are at least four disadvantages for society or the environment of using technology?

s. What are at least four technological devices that you rely on on a day-to-day basis?

t. Is an automobile technology or a product of technology?

u. What should be the role of scientists in making use of technology?

v. What are the worst ways people can put technology to use?

w. What are your worst fears about the ways science could be used?

x. What are your worst fears about the ways technology could be used?

y. If you could contribute to one scientific or technological breakthrough in your lifetime, what would you hope that breakthrough would be?

YE MAY KEEP YOUR PET ROCKS

The U.S. Department of the Interior has announced proposed regulations governing the importation of wildlife that are injurious to human beings, forests, agriculture, horticulture, and native wildlife in this country. A department spokesperson reported the following:

> On the average, eight people a year are bitten by imported poisonous snakes. In just one year, exotic pets seriously injured 190 people in New York City alone. Small turtles in pet stores are estimated to cause more than 40,000 cases of salmonella poisoning a year. Even though this type of food poisoning is usually curable when treated in time, it is still dangerous. Imported monkeys can infect humans with a disease of the liver called hepatitis. Tuberculosis, a disease affecting the lungs, can also be transmitted by monkeys. Venomous fish set free in our

nation's lakes and rivers have poisoned and paralyzed people. Newcastle disease, brought in by parrots and mynah birds, affects the respiratory and nervous systems. In 1973, this disease killed over 11 million chickens in California.

The proposed regulation would stop the shipment of injurious wildlife into the United States. The Secretary of the Interior could allow some animals to be imported for scientific, education, zoological, or medical purposes. The proposal includes a list of "low risk" wildlife. This means all species not listed as "low risk" would be prohibited from importation except under a strict permit system and for the reasons indicated above.

Should these regulations be adopted, they would significantly impact the U.S. pet industry. The ornamental aquarium fish and accessories trade is expected to be the hardest hit. These regulations would reduce importation of birds (mostly parrot and mynah birds) by about 50 percent, mammals (primarily monkeys) by 45 percent, and reptiles (especially snakes) by 95 percent. Such reductions may close many pet stores. Department stores with pet sections would likely close their pet sale divisions.

Many of the animals banned for sale to the public would be allowed to enter the country for scientific research purposes. A different set of laws would be passed to regulate the importation, sale, and use of animals for scientific studies.

1. How would the department spokesperson define *wildlife?*

2. What government agency is seeking to restrict the importation of wildlife?

3. What have been three negative consequences of bringing unauthorized wildlife into the United States?
   a.

   b.

   c.

4. In what two ways might imported wildlife be considered pollutants?
   a.

   b.

5. Besides those in the announcement, what are three ways these government regulations might affect the American pet industry?
   a.

   b.

   c.

6. The announcement included a number of details that reflect the work of scientists. What are three items in the statement that reflect the findings of scientists?
   a.

   b.

   c.

7. The Department of the Interior actively seeks to restrict wildlife considered dangerous to the environment. What are two reasons this is a good policy for a government agency to enforce?
   a.

   b.

8. Suppose you were waiting for an imported pet, but discovered these new regulations prohibited it from being brought into this nation. What words would best describe your emotions?

   What are the reasons you would feel this way?

9. Suppose the Department of the Interior allowed scientists to import animals that citizens of the United States were banned from importing as personal pets. What words would best describe your feelings toward those scientists?

REVIEW AND REFLECTION QUESTIONS

Suggested follow-up questions to focus and guide inquiry and learning.

- **a.** Suppose the Department of the Interior asked you to list ways you could help preserve the environment. What four policies would you suggest?
- **b.** If you were a scientist in this situation, how might you react to these new regulations?
- **c.** What roles could scientists play to prevent the negative effects of importing animals?
- **d.** What roles have scientists probably played in trying to prevent negative consequences of imported animals?
- **e.** To what extent should scientists become involved in increasing the types of animals imported into a country? Into this country?
- **f.** In what ways might the regulations discussed in this reading be the results of scientific studies?
- **g.** If an animal is labeled as "wildlife," what does this say about that animal?
- **h.** If an animal is labeled as a "pet," what does this say about that animal?
- **i.** How have scientists helped convert many forms of wildlife to pets?
- **j.** What types of animals do scientists use in their experiments to better human life?
- **k.** To what extent would it be fair to Americans to allow animals to be imported only for scientific research?
- **l.** What is an *exotic* animal?
- **m.** To what extent should scientists be allowed to do research on animals the general public can't have for pets?
- **n.** How might the government use technology to stop the importation of dangerous animals into a country?
- **o.** What are three ways that humans might use animals to acquire new knowledge?

EPISODE **31**

OUT OF FRUSTRATION

Rosa was troubled. For several years she had watched her teachers do a number of science experiments from the science book. She had tried to understand the purpose of the experiments and how they were tied to the topics being studied. She could comprehend the information she was studying. She could usually follow the way the experiments were done. When Rosa would ask for a better description or explanation of what was done, her teachers often became upset with her. She liked to make her own predictions about what might happen, but her teachers usually told the class what the results from the experiments should be. When she was being told what to expect, Rosa would usually cover her ears. When she did this, her friends thought she was strange. Rosa was always excited when the results were different than what she had been told. But when she offered these as explanations to her classmates, they were not interested.

OUT OF FRUSTRATION

One day, after Rosa's teacher told her that her explanations during class were unimportant, Roger caught up with Rosa as she was walking home. He told her he too wanted a better scientific understanding of the experiments and their results. He had been afraid to speak up as she had done. They decided they had to find someone to talk to in order to get more information about what science is and how it inquires into scientific problems. They had heard about another teacher in their school who knew a lot about how scientists think and learn.

They made an appointment to meet with the teacher. They said it was out of frustration they had come to see her. After ten minutes, the teacher handed them the following information on a computer printout:

> Science has many definitions. For some, science is an organized accumulation or collection of information (such as facts, observed data, and interpretations) describing what seems to occur in nature and how these occurrences or things are connected to a specific topic or idea. Science is also seen as a systematic method, a controlled way for any person to acquire enough information to solve a particular problem. For others, science is a way to describe, analyze, test, and explain what happens in the world and to predict what may likely happen.
>
> Three major aims of science are description, explanation, and prediction. *Description* is detailed information on the observable features or characteristics of any thing, event, or phenomenon. *Descriptive information* is basic and cannot be overlooked if inquiry is to go beyond the fact-gathering stage. *Explanations* involve using reasons, causes, concepts, and principles to interpret specific things or phenomenon. *Predictions* are educated guesses or hypotheses made by applying particular information that make correct predictions likely.
>
> *Empirical science* involves finding valid and reliable principles or laws using standards or ideals to verify the adequacy of empirical scientific knowledge. One ideal asks that people be as free as possible from personal or cultural bias regarding the topic to be researched. People must be objective, or unbiased, because they must logically confirm or not confirm the conclusions obtained by using various methods of investigation.
>
> A second ideal enables people to distinguish between opinions (sometimes even superstitions) and verifiable information about things. When based on techniques of experimentation and careful analysis of the results, empirical

scientific information should be reliable (that is, dependable). This information is said to describe exactly what seemed to happen under certain conditions.

A third ideal states that people need to extend and amplify their senses by using tools such as telescopes, Geiger counters, microscopes, and all other instruments of investigation so they don't limit their scope of information gathering.

A fourth ideal requires that the concepts used within scientific methods of investigation and interpretations are precisely defined. General impressions may be used to make estimates, but, to increase objectivity, techniques of precise measurement and definitions must be used.

A fifth ideal states that people seek to construct a connected, systematic description and explanation of what occurs in the world—and why it occurs. The results of their inquiry is not a collection of miscellaneous bits of information. Descriptions might be in the form of detailed and precise definitions, charts, diagrams, and classification systems. Explanations and predictions may include concepts or procedures as well as scientific principles, theories, and hypotheses. With precise and reliable information and methods of systematic investigation, people can better judge whether a phenomenon or event actually occurred.

A scientist is expected to always be ready to revise or abandon an original conclusion should there be evidence gained from the comprehensive study and investigation that would make this revision necessary. Unfortunately many scientists are not willing to change their minds after contrary findings and interpretations are presented. Decisions not to change their minds are a reminder that scientists, even the very best scientists, are human beings.

Although many people think there is, *there is no single scientific method.* Empirical science is not the only way of "doing" science. There are alternative methods scientists use in their investigations. Sometimes scientists working on the same problem or on the same scientific team use very different methods of inquiry. *Do not make the mistake of thinking that only one method of science exists.*

Learning science is not memorizing and recalling a lot of definitions, terms, and description of things. It is learning how to think and act in particular ways. Scientific thinking includes making judgments and interpreting events in the world.

It involves using the methods of systematic investigation to describe and explain what exists—and why something happens as it does. Sometimes the results of these methods can be used to make reasoned predictions.

Roger and Rosa were excited by what they just found out about science and science investigators. They realized how important it is to question why certain things happen as they do in experiments. They also learned science has many accepted meanings.

Now that you have read this story about Rosa and Roger write answers to the questions below. Try to do this first without referring back to the story. Look back only when you absolutely must. Be ready to give reasons for your answers.

1. According to the reading, what is *science?*

2. What kinds of people can be scientists?

3. Imagine that you saw someone "doing" science. What would you observe this scientist doing?

4. In your own words, what are five major ideals of scientists?
   a.

   b.

   c.

   d.

   e.

5. Which of these ideals of scientists do you like best?

   What scientific reasons do you have for this choice?
   a.

   b.

   c.

6. Which of these ideals of scientists would make science exciting for you?

7. If you were in science classes such as those described in this story, what reaction would you have toward the teacher as a promoter of "scientific thinking"?

   What reasons would you have for this reaction?

8. In what ways is the view of science in this story the same as what you thought science was before you read this story?

9. In what ways is the view of science in this story different from what you thought science was before you read this story?

REVIEW AND REFLECTION QUESTIONS

Suggested follow-up questions to focus and guide inquiry and learning.

a. What is a *method of science?*

b. What is a *scientific fact?*

c. What is a *scientific explanation?*

d. What is it about a scientific explanation that makes it scientific?

e. What is the major difference between *empirical science* and *science?* Of these definitions, which should be accepted by most people in this society? Of these definitions and ideals, which is closest to what you thought science was before you read the story?

f. Why would scientists want the information they collect to be reliable?

g. What does a scientist mean when he or she talks about being *objective?*

h. Suppose a scientist said scientists should be subjective. If every scientist were very subjective in his or her work, would this be good or bad for science?

i. According to this story, to what extent are all scientific explanations able to be used as predictions?

j. What might lead a scientist to refuse to accept interpretations and findings different from the interpretations he or she already had about something?

k. If scientists refused to seriously consider and accept findings different from what they already believed, what would this say about scientists? About science?

l. Would it be good or bad for a society to have scientists who met these ideals?

m. Of *description, explanation,* and *prediction,* which is the most important for scientists to do? The least important? The most exciting for you?

n. When scientists are involved in scientific investigation, what are at least three emotions they likely feel?

o. What areas of science would be the most exciting for you? The most boring for you?

p. What would lead you to become a scientist?

q. What was the main theme of this story?

r. If you forgot everything about this story except one thing, what would that one thing be?

s. Imagine you were Rosa in this story. After you read about science, what words would best describe your feelings about your past behaviors in your science class?

DIRTY SNOWBALLS

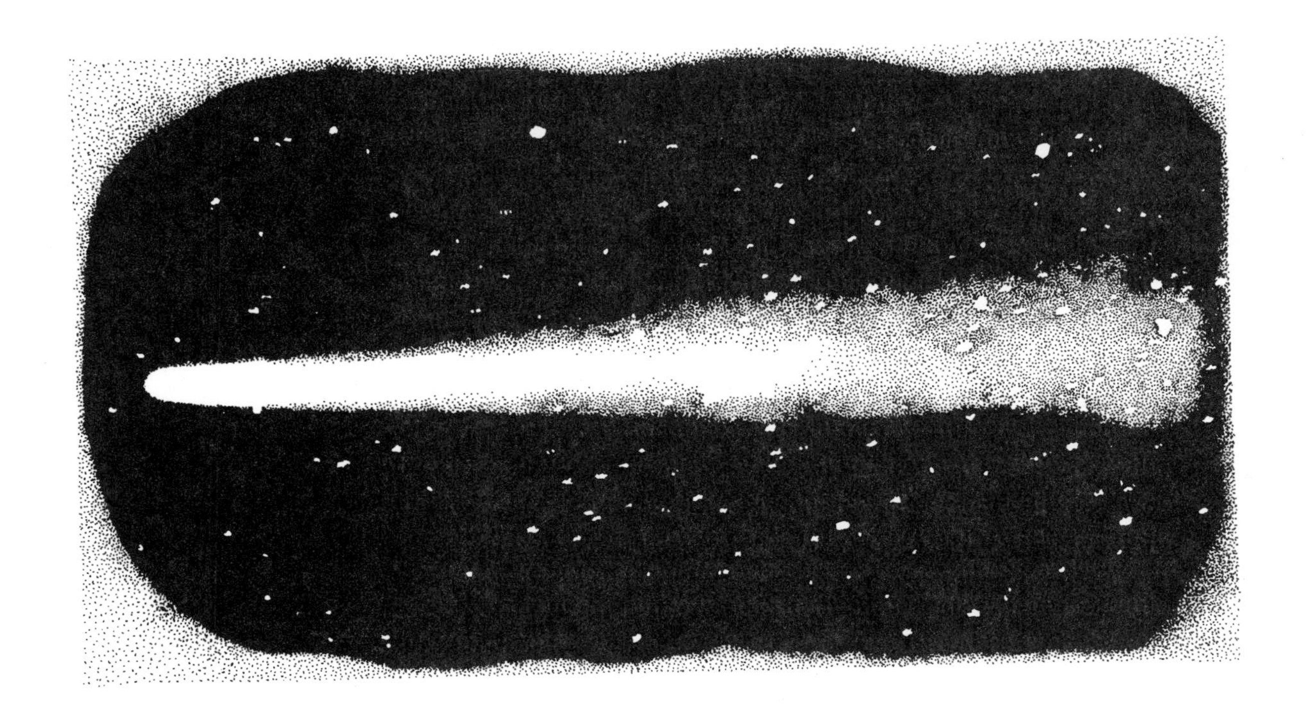

A University of Arizona planetary scientist, Uwe Fink, made a recent discovery that may force scientists to rethink the origins of comets. He made an announcement after analyzing the ingredients in a comet that passed close to Earth in early 1989. His analysis may also lead to rethinking our theories on the origins of our solar system.

Dr. Fink observed Comet Yanaka 1988r with the University's 61-inch telescope on Bigelow Mountain, northeast of Tucson. He studied the gases and dust pouring off the comet's surface as it rounded the sun on the way back toward the outer planets of the solar system. Dr. Fink has analyzed the chemical composition of over 20 comets since 1986.

A typical comet is made of ice and dust. For this reason, comets are often called "dirty snowballs." They contain a great deal of frozen water. They also contain frozen carbon dioxide, carbon monoxide, and ammonia. The frozen gases evaporate as the comet warms, releasing both carbon and silicate dust. These escaping gases contribute to the tail of the comet.

Comet Yanaka 1988r appears to contain much less carbon than any other comet Fink has studied. Its composition is vastly different from what was expected. This difference means all comets in our solar system did not originate from the same source.

Currently two theories of comet origins are most accepted by astronomers. The leading theory states that comets were created near Uranus and Neptune about 4.5 million years ago. They were formed as gas and dust surrounding the early sun began to condense to form planets. Some of the comets spun off into the Oort Cloud, a group of perhaps a billion comets far beyond the known planets. From time to time, some of these comets are pulled from their positions by the gravitational pull of a nearby star. As they move out of position, they move toward our solar system and the sun.

In this theory, called the Oort Cloud Theory, the assumption is that all comets were formed at the same time as our solar system. They are assumed to have been formed in the same region of space and of the same materials. The theory assumes that the region where these comets were formed contained a fairly uniform distribution of gases, dust, and water.

The second leading theory suggests that comets formed in a cloud of gas and dust outside our solar system. Later these comets were "captured" when our solar system passed through the assortment of comets. Gradually, these comets are pulled in toward the center of our solar system. However, this theory also assumes that the cloud within which the comets were formed had a even distribution of gases and dust.

Yanaka 1988r was expected to have the same or almost the same chemical and dust composition as all the other comets that have been analyzed. Because of its unique composition, Yanaka's origin is not likely to be the same as that of the other comets that have been studied.

1. In their research on comets, what technology might astronomers such as Ewe Fink use?

2. In your own words, what major idea is expressed in this article?

3. In this article, what are the three most important facts about comets?
   a.

   b.

   c.

4. In your own words, what is the Oort Cloud Theory of comets?

5. Based upon the data in this article, in explaining the origins of comets, how adequate is the Oort Cloud Theory of comets?

6. How valuable will research on comets be to your society and you?

7. What does this article reveal about the work of scientists relative to scientific theories?

8. What is the value of studying the composition and origins of comets?

9. If scientists ever do find the origins of comets, how will this news affect how you think and live?

10. Given the problems and needs of modern society, why should people in our society continue to study space objects such as comets?

REVIEW AND REFLECTION QUESTIONS

Suggested follow-up questions to focus and guide inquiry and learning.

a. In doing their work, what are three major jobs astronomers need to do?

b. What is an *observatory?*

c. In your own words, what did Uwe Fink discover about Comet Yanaka 1988r?

d. Columbus is said to have "discovered" the land that later became the Americas. Dr. Fink is said to have "discovered" the composition of Comet Yanaka 1988r. In what ways were these discoveries identical?

e. What does this article reveal about how permanent theories are in certain areas of science?

f. In all fields of science, what are the roles of theories?

g. In your own words, what is a *theory?*

h. What are at least three good reasons a society should pay people to study the origins and composition of comets?

i. If you could discover something about space, what would you want to discover?

j. If you could look into space through the most powerful telescope on Earth, what object or objects would you most like to observe?

k. If you were to look through such a telescope, would you be a scientist at that moment?

l. A telescope might be viewed as a scientific tool. What are the names of other tools astronomers use?

m. Some telescopes cost millions of dollars to design, build, and use. What are the reasons people would spend millions of dollars to study "dirty snowballs"?

n. Besides planets, the moon, the sun, and comets, what are at least three other things in space astronomers might study?

o. What is the most important object in space that astronomers should study?

p. Based upon articles such as this one, one might conclude that not all comets were formed at the same time or even in the same region of space. In what ways might this conclusion be a theory?

q. What does it mean to "discover" something?

r. What are three ways scientists might use technology to find out more about comets?

RAIN OR SHINE?

PHOENIX

Today: Mostly sunny, high 92
Tonight: Clear, windy, low 66
Tomorrow: Partly cloudy, high 90, low 63

EXTENDED FORECAST

Monday: Partly cloudy, high 91, low 67
Tuesday: Sunny, high 93, low 68

YESTERDAY

High 91, low 67
Rainfall: 0.00
Rainfall to date: 10.48"
+ or – rainfall to date: +4.62"
High/low relative humidty: 33%/15%
Barometric pressure at noon: 29.855
Record high this date: 97 in 1990
Record low this date: 42 in 1956
Normal high/low: 88/60
Average wind speed: 3.5 mph
Peak wind gust: 15 mph from west
High/low dew point: 44/33

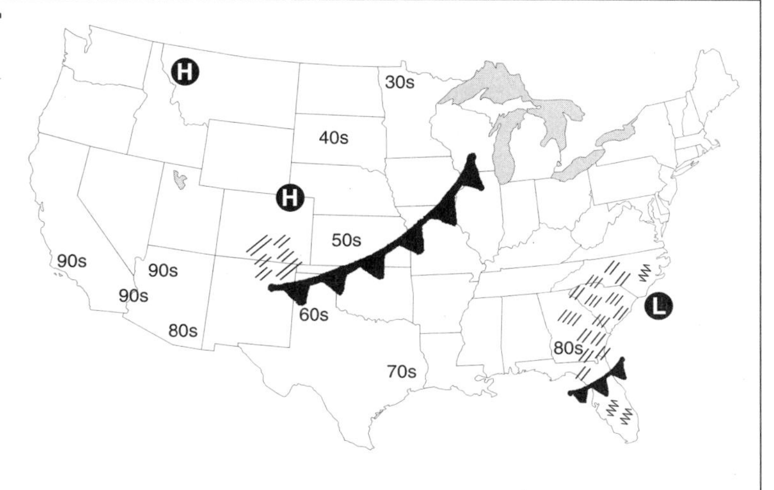

SUN INTENSITY

Minutes in the sun today to redden untanned and unprotected skin
9 a.m.: 60
Noon: 36
3 p.m.: 48

POLLEN/MOLD

Average for samples taken at stations throughout Maricopa County for the 24-hour period ending at 5 p.m. yesterday. Figures represent the number of parts found in a cubic meter of air.
Very high (51+): Cladosporium, 239, Altermaria, 135
High (26–50): Smuts, 35
Medium (11–25): Misc. molds, 24, Bipolaris, 17
Low (5–10): Ragweed, 9; Leptospherulina, 5

WEATHER ACROSS THE NATION

	Friday	Today	Sunday
	Hi Lo Prec	*Hi Lo Sky*	*Hi Lo Sky*
Albany	60 39	60 33 pc	52 43 r
Anchorage	33 30 .03	32 27 sn	34 23 c
Boston	65 50	52 43 c	52 41 r
Dallas	75 46	72 48 pc	62 45 sh
Miami	86 75 .07	87 71 ts	86 72 pc
Seatle	61 41	63 44 s	62 39 s

s–sunny; c–cloudy; pc–partly cloudy; r–rain; sh–showers; ts–thunderstorms; sn–snow; sf–snow flurries; bl–blizzard; ic–ice; w–windy; cl–clear; nr–no report

FOR AVIATORS

VFR conditions across the state today with some turbulence expected in the northeast and east central. Winds will be northwesterly at 7 to 11 mph above 7000 feet. Below 7000 feet, winds will be north at 8 to 16 mph.

SUN AND MOON

Sunset today	5:42 p.m.
Sunrise tomorrow	6:41 a.m.
Moon rises tonight	3:39 p.m.
Moon sets tomorrow	4:46 a.m.

WORLD WEATHER

	Friday	Today
	Hi Lo Prec	*Hi Lo Sky*
Cairo	68 53	67 52 pc
Paris	43 36 .09	44 37 cl
Perth	70 46	71 48 s

LAWN WATER NEEDS

0.3" is needed on your Bermuda grass if you last watered three days ago, unless it has rained since.

RAIN OR SHINE?

The information above might be found in a typical weather report in a newspaper. The questions below are directed toward this report.

1. What are five specific areas of information included in this report?
   a.

   b.

   c.

   d.

   e.

2. What area of information in this report, and in others like it, is *most* important for you?

3. What area of information in this report, and in others like it, is *least* important to you?

4. Suppose the data above came from today's newspaper. From these data, what will today's weather be in your town?

5. Suppose a friend said the only thing a weather report can say is whether it will rain in a particular area. Suppose you lived in Phoenix and the above information was printed in a Phoenix newspaper. How would you respond to your friend's statement?

6. By specific name, what are all the areas in this report that reflect the work of scientists?

7. What specific areas in this report could be done *without* the use of science?

8. What are all the areas in this report that reflect the use of technology?

9. What specific areas in this report could be done *without* the use of technology?

10. If people used weather reports to decide what they will or won't do on a particular day, to what extent would these people be using the results of science and technology to influence their lives?

REVIEW AND REFLECTION QUESTIONS

Suggested follow-up questions to focus and guide inquiry and learning.

a. What is the definition of *weather?*

b. What is the definition of *climate?*

c. What is the most important way in which weather differs from climate?

d. What specific parts of the weather data in this report do not involve scientific ideas or principles?

e. For what reasons would this weather report be considered a scientific document?

f. Look at a weather report in your local paper. What specific weather information does it contain?

g. In what specific ways is the weather information in this report identical to information in your local paper?

h. In what specific ways is the weather information in this report different from information in your local paper?

i. For what reasons would scientists be interested in the weather? In climate?

j. What makes certain of the information in this weather report "scientific"?

k. What are three ways you could collect scientific information about the weather in your area?

l. For what reasons would a thermometer be a "scientific instrument"?

m. If you used a thermometer to find the temperature outside your school building or home, would you be a scientist?

n. How would you have to use a thermometer to use it "scientifically"?

o. In what ways is a thermometer an example of technology?

p. Is it possible for a thing to be both a scientific instrument and an example of technology? If so, in what situations is an example of technology not a scientific instrument?

q. Suppose a next-door neighbor made a daily record of the temperature, rainfall, and humidity in his or her back yard. In what specific ways would this person be "doing" or not "doing" science?

r. In the above situation, what would your neighbor have to do to make sure his or her activities were consistent with "doing" science?

s. How often do you use the weather report in your local newspaper?

t. If you used such weather information to make decisions about your activities, to what extent would you be making use of scientific findings and the results of the use of technology?

APPENDIX

Optional Lesson Plans

This Appendix contains optional lesson plans for 20 of the 33 episodes. The lesson plans are in an easy-to-follow, practical format and require minimal teacher or student preparation. They contain many helpful suggestions on how you can use the episodes. Use the lesson plans as stated or revise them; they may complement the plans you have already developed. Feel free to modify them or to use them in learning centers as guides for students.

Study each episode, and then revise or supplement the lesson plan to meet your own needs. The episodes may be used for a great many situations besides those outlined in these lesson plans. Other uses may be far more powerful than those suggested here. It would be impossible to develop a set of lesson plans for every possible use of each episode. We urge you to quickly move away from these plans to incorporate the episodes into ongoing units and activities you have already planned.

The teacher preparation suggestions under Description and the Lesson Plan Suggestions are not required parts of an episode. They suggest ways you may enrich, extend, and individualize instruction to fit the needs of particular students and classroom situations. Nearly all of the episodes are readily adaptable for use in learning centers and cooperative-learning groups.

Note: You may want to distribute the Individual Decision Sheet only after students have had sufficient time to answer the questions on the Predecision Task Sheets. When these two sheets are passed out together, a number of students will probably not complete the Predecision Task Sheets and will hurry on to the Individual Decision Sheet. Also, one copy of the Group Decision Sheet is given to each group. When students work in groups, you may want to wait and pass out the Group Decision Sheet only after all students have had time to complete the Individual Decision Sheet.

Each lesson plan follows the format outlined below.

Content Focus: The major subject matter area or event covered by the episode.

Episode Title: A catchy heading.

Duration: The suggested or approximate amount of time the episode should take. This is provided for the your convenience in planning.

Purpose: The major goal or aim of the episode. This may serve as the rationale for using the episode.

Decision Strategy: The decision strategy or strategies students will be required to use to complete the episode.

Description: A description of the episode, including steps and procedures you may want to follow to help students obtain the appropriate and necessary background to adequately prepare for the activity.

Content Objectives: The subject matter content outcomes expected to result from appropriate use of the episode.

Process Objectives: Decision-making and problem-solving processes students will be using as they respond to the activity.

Vocabulary: New and/or important words or terms students will be using in the episode. Should students be unfamiliar with these terms, a short definition session may be needed before the activity begins.

Materials: Resources and aids needed to complete the episode. In most cases, this section only informs you to reproduce copies of the pages that make up the particular episode.

Lesson Plan Suggestions: Decisions and steps you may want to make before, during, or after using an episode to individualize or enrich it, or to help students connect it to other things they are studying.

EPISODE 1

Content Focus: Herbivorous dinosaurs

Episode Title: My Favorite Monster

Duration: 1 class period

Purpose: To help students comprehend similarities and differences among herbivorous dinosaurs

Decision Strategy: Forced choice

Description: The story describes an expedition to Africa to prove or disprove rumors of a large, dinosaurlike creature roaming the jungles. Six types of vegetarian dinosaurs are described. Students should individually complete Predecision Task Sheets 1 and 2 after they read the story. They meet in a small group to discuss their responses, and then write their decisions on the Individual Decision Sheet. The Group Decision Sheet gives students the opportunity to consider collectively the scientific need to investigate prehistoric life forms.

Content Objectives

- Students will state the names of six herbivorous dinosaurs.
- Students will be able to describe the six herbivorous dinosaurs based on scientific information from fossils and speculation from today's reptiles.
- Students will state scientific facts and speculations about the habits and routines of some prehistoric dinosaurs.

Process Objectives

- Students will weigh given alternatives and choose only one of the alternatives.
- Students will consider logical reasons in favor of each of the possible choices and their decisions.
- Students will consider the consequences involved in choosing among alternatives.

Vocabulary: amphibian, apotosaurus, brachiosaurus, camarasaurus, camouflage, carnivorous, diplodocus, expedition, ganglia, herbivorous, plateosaurus, prehistoric, pygmy, vegetarian, yaleosaurus

Materials: Hand out copies of the story, Predecision Task Sheets, and Decision Sheets.

Lesson Plan Suggestions

- Decide if and when students will study prehistoric vegetarians.
- Decide if and when students will study and make lists of the equivalencies, similarities, and differences between herbivorous and carnivorous dinosaurs.
- Decide if and when students will study other forms of life during the Mesozoic Age.
- Decide if and when students will visit a science museum that has exhibits of prehistoric life.
- Decide if and when students will make a sketch of the "mystery monster."
- Decide if and when students will examine descriptions and reports of dinosaur fossils.
- Decide how students will be organized into small groups to reach a consensus.
- Decide which Review and Reflection Questions to ask to focus the general discussion following the story.
- Decide how you will incorporate the concepts from this episode into the next activity you use.

EPISODE 3

Content Focus: Community conservation projects

Episode Title: Choose or Lose

Duration: 1 class period

Purpose: To help students understand ways to protect and enjoy the environment

Decision Strategy: Forced choice

Description: This episode is a small-group decision-making activity in which students must reach a consensus as members of a city council deciding on one of three proposed projects. Environmental considerations are dealt with on the optional Predecision Task Sheet. Each student reads the story carefully and completes an Individual Decision Sheet. Then students meet in groups to reach a consensus.

Content Objectives

- Students will comprehend how some projects might protect the environment.
- Students will study a number of possible benefits of three environmental projects.
- Students will define *environment* and consider the implications of actions to protect or disturb it.

Process Objectives

- Students will weigh given alternatives and choose only one of the alternatives.
- Students will consider the consequences involved in choosing among particular alternatives.
- Students will invent and consider logical reasons that support their decision.

Vocabulary: conserve, environmentalist, wildlife refuge

Materials: Hand out copies of the story, Predecision Task Sheets, and Decision Sheets to each student. Prepare the Group Decision Sheet if you want students to work towards a group decision.

Lesson Plan Suggestions

- Decide if and when students will examine existing city projects that preserve and protect the environment.
- Decide if and when students will discuss the building of roads and parking lots as possible pollutants to the environment.
- Decide when and how students will define *conservation.*
- Decide when and how students will complete the Predecision Task Sheet.
- Decide if and when students will discuss city projects the finances involved in projects such as those proposed in the story.
- Decide how students will be organized into small groups to reach a consensus.
- Decide which Review and Reflection Questions to ask to focus the general discussion following the story.
- Decide how you will incorporate the content and concepts from this episode into the next activity you use.

EPISODE 5

Content Focus: Scientific data on psychological stages of death and dying

Episode Title: The Circle

Duration: 1 or 2 class periods

Purpose: To help students comprehend facts relating to a prolonged illness and the complications of scientific research on dying

Decision Strategy: Forced choice

Description: The story deals with whether to tell family members and/or the dying individual they are dying. Each student reads the story carefully and completes the Predecision Task Sheet. Then students meet in small groups to reach and then record their consensus decision on the Decision Sheets.

Content Objectives

- Students will comprehend some of the positive and negative consequences of telling a person they are dying.
- Students will define dying and the stages of dying.
- Students will consider some ways people think when they know they or a loved one is going to die.

Process Objectives

- Students will weigh given alternatives and choose only one of the alternatives.
- Students will consider the consequences involved in choosing among alternatives.
- Students will give reasons to justify their decision.

Vocabulary: conscious, death, dying, grieving, hospice home care, Elisabeth Kübler-Ross, physical therapist, rejection

Materials: Hand out copies of the story, Predecision Task Sheets, and the Individual and Group Decision Sheets.

Lesson Plan Suggestions

- Decide if and when students will discuss the possible effects of brain damage.
- Decide if and when students will discuss definitions of death and dying.
- Decide if and when students will examine the changes that might take place among family members following a serious accident.
- Decide how students will be organized into small groups to share their ideas about the story.
- Decide when and how terms will be defined by students.
- Decide which Review and Reflection Questions to ask to focus the general discussion following the story.
- Decide how you will incorporated the concepts and content from this episode into the next activity you use.

EPISODE 6

Content Focus: Air pollution and the health of employees

Episode Title: No Deep Breaths

Duration: 1 class period

Purpose: Students will understand how poor health precautions could lead to demands for alternative working conditions

Decision Strategy: Forced choice

Description: The story deals with poor working conditions in a textile mill. Various policies are outlined that might bring about changes for the employees and managers. The episode requires students to make an individual decision. Following each of the Predecision Task Sheets, students find out what other students think about the problem and clarify terms.

Content Objectives

- Students will comprehend facts concerning poor working conditions in a mill.
- Students will consider how individuals who are concerned with a particular working condition can influence the passage of improved company policies.

Process Objectives

- Students will weigh the alternatives given to them in a situation.
- Students will consider the consequences of those alternatives.
- Students will devise a logical argument in a favor of their decision.

Vocabulary: boycott, brown-lung disease, minimum wage, monotony, sabotage, slowdown, strike, unskilled laborer, violence

Materials: Hand out copies of the story, Predecision Task Sheets, and Decision Sheets.

Lesson Plan Suggestions

- Decide if and when students will examine the history of labor reforms in mills.
- Decide if and when students will investigate some of the measures workers can take to improve their working condition.
- Decide if and when students will discuss poor working conditions that have existed in mills.
- Decide if and when students will study the pros and cons of labor unions or labor reforms.
- Decide when and how students will define the terms used in the episode.
- Decide which Review and Reflection Questions to ask to focus the general discussion following the story.
- Decide how you will incorporate the concepts from this episode into the next activity you use.

EPISODE 7

Content Focus: Hiking and rappelling safety precautions

Episode Title: Cliff Hanger

Duration: 1 class period

Purpose: To help students understand the need for safety precautions during all phases of a sport or activity

Decision Strategy: Forced choice

Description: The story deals with two friends who, after hiking all day, realize they cannot get back to their campsite before dark. Hiking and rappelling skills are discussed. The story provides the necessary information for a decision. Each student reads the story carefully, completes the Predecision Task Sheets, and records his or her individual decision on the Individual Decision Sheet. Then students meet in groups to reach consensus.

Content Objectives

- Students will comprehend facts and rules about hiking and rappelling.
- Students will comprehend the need to follow the safety rules of a sport.
- Students will define *safety.*

Process Objectives

- Students will weigh given alternatives and choose only one of the alternatives.
- Students will consider the consequences involved in choosing among alternatives.
- Students will consider logical reasons in favor of their decisions.

Vocabulary: carabinier, descend, hiking, rappelling, results

Materials: Hand out copies of the story, Predecision Task Sheets, and Decision Sheets.

Lesson Plan Suggestions

- Decide if and when students will discuss the activities of hiking and rappelling.
- Decide if and when students will discuss various articles on and pictures of hiking and rappelling.
- Decide how and when terms will be defined by students.
- Decide if and when students will discuss their personal definitions of safety.
- Decide if and when students will discuss the reasons for safety rules and the need to follow the rules of a sport.
- Decide which Review and Reflection Questions to ask to focus the general discussion following the story.
- Decide how you will incorporate the concepts from this episode into the next activity you use.

EPISODE 8

Content Focus: Objects in the solar system

Episode Title: Heavenly Bodies

Duration: 2 class periods

Purpose: To help students understand eight objects in space and how information about them could be helpful to human life and Earth

Decision Strategy: Rank order

Description: The story concerns eight space missions that could be conducted to better understand the solar systems' past and possible future. Predecision Task Sheet 1 has students collect facts and data about the different space objects and what we could learn from each. Predecision Task Sheet 2 stresses the concept of space exploration from the perspective of a scientist. The Decision Sheet asks students to rank order the space missions according to their possible value to Earth and the future of humanity.

Content Objectives

- Students will comprehend and analyze critical pieces of information about eight space objects.
- Students will comprehend the activities of an astronomer as a scientist.
- Students will reflect on the value of a number of space projects and their likely benefits to people and life on Earth as well as to various fields of science and technology.

Process Objectives

- Students will weigh alternatives appropriate to the situation.
- Students will consider the long-term consequences of those alternatives.
- Students will rank alternatives from the most valuable to the least valuable.
- Students will support their decisions and ranks with logical reasons.

Vocabulary: asteroid, comet, glacier, gravitational pull, mission, orbit, planet, rotates, solar winds

Materials: Hand out copies of the story, Predecision Task Sheets, and Decision Sheets.

Lesson Plan Suggestions

- Decide if and when students will examine a model of the solar system.
- Decide if and when students will be allowed to use reference books and materials on the planets and sun.
- Decide if and when students will study the work of astronomers.
- Decide if and when students will make a time line of previous space missions.
- Decide whether students will write letters to Congress and/or the President to encourage or discourage space missions.
- Decide whether students will search for and study other space objects that could be investigated by space missions.
- Decide if and how students will be selected and assigned to the small groups you may use for the episode.
- Decide whether students will complete Predecision Task Sheet 2 as part of the episode.
- Decide which Review and Reflection Questions to ask to focus the general discussion following the story.
- Decide how this episode will be followed by the next activity you use.

EPISODE 9

Content Focus: Recording scientific observations and experiments

Episode Title: Oops! The One Time I Forgot!

Duration: 1 class period

Purpose: To help students understand many of the reasons scientists are encouraged to keep accurate records of their activities

Decision Strategy: Rank order

Description: The episode stresses the role of record keeping in conducting scientific experiments and observations. It points out some problems when such records are not complete or accurate. During this episode, students rate several alternative solutions. Predecision Task Sheet 1 has students review information about records scientists keep and review terms. Predecision Task Sheet 2 has students individually consider the possible choices and then share their responses with others. Then students record their rankings on the Decision Sheets.

Content Objectives

- Students will consider steps involved in accurately charting lab procedures.
- Students will comprehend positive and negative results of keeping accurate records of one's scientific experiments and observations.

Process Objectives

- Students will consider the positive and negative consequences of different courses of action.
- Students will rank the alternatives according to contextual and personal criteria.

Vocabulary: chart, chemicals, discovery, experiment, method, microscope, observation, organism, record, science, scientist

Materials: Hand out copies of the story, Predecision Task Sheets, and Decision Sheets.

Lesson Plan Suggestions

- Decide if and when students will investigate the scientific methods represented in writing lab reports.
- Decide when and how the terms used in the episode will be defined.
- Decide if and when students will discuss the differences and connections between scientific experiments and scientific observations.
- Decide if and when students will discuss the differences between science fiction literature and movies as opposed to scientific reality.
- Decide which Review and Reflection Questions to ask to focus the general discussion following the story.
- Decide how you will incorporate the decisions from this episode into the next activity you use.

EPISODE 12

Content Focus: Scarcity among natural resources

Episode Title: Those Pesky Pesticides

Duration: 1 class period

Purpose: To help students understand alternatives to be considered should a resource suddenly become scarce

Decision Strategy: Rank order

Description: This episode is designed to be completed by a small group, although an individual could complete it alone. The story deals with a successful company's dilemma in conserving natural resources. Predecision Task Sheets 1 and 2 give students an opportunity to review information about pesticides and define terms. Students will individually rank order choices between saving a forest and saving a lumber company. Then they meet in a small group, reach a consensus decision within the group, and record their group's consensus rankings.

Content Objectives

- Students will apply the concepts of *scarcity* and *resource.*
- Students will consider alternative solutions should trees become a scarce resource to lumber companies.
- Students will comprehend possible consequences of pesticide use in forests.
- Students will state different uses of hardwoods and softwoods.

Process Objectives

- Students will weigh alternatives appropriate to the given situation.
- Students will consider the consequences of those alternatives.
- Students will rank alternatives from the best to the worst.
- Students will devise a logical argument in favor of their decision.

Vocabulary: ecologist, infest, marginal land, pesticide

Materials: Hand out copies of the story, Predecision Task Sheets, and Decision Sheets.

Lesson Plan Suggestions

- Decide if and when students will explore the different uses of hardwoods and softwoods.
- Decide if and when students will examine samples of hardwood and softwood.
- Decide if and when students will study the economics required for starting and maintaining a large forest-industry company.
- Decide if and when students will study planting and irrigation methods for growing trees.
- Decide if and when students will discuss other uses and results of pesticides.
- Decide how students will be organized into small groups to share their ideas about the story.
- Decide when and how terms will be defined by students.
- Decide which Review and Reflection Questions to ask to focus the general discussion following the story.
- Decide how you will incorporate the concepts from this episode into the next activity you use.

EPISODE 13

Content Focus: Balance between community needs and environmental concerns

Episode Title: Representing the People

Duration: 1 class period

Purpose: To help students analyze the environmental balance between nature and a growing business community

Decision Strategy: Rank order

Description: The story concerns environmental problems encountered as an area's economy is boosted by tourism. Five legislative bills that might protect the environment are described. Each student should read the story carefully, then use the Predecision Task Sheets to consider the consequences of the bills prior to making a final decision. Students may meet in small groups to discuss the issue and clarify terms. Students record their individual ranks on their Individual Decision Sheet. Then students meet in groups to complete the Group Decision Sheet.

Content Objectives

- Students will comprehend five actions people might take to protect the environment.
- Students will describe how tourism may benefit and disturb a community and its environment.
- Students will comprehend how political power may be accessed by the public, such as people in environmental action groups.
- Students will consider the responsibilities of elected officials.

Process Objectives

- Students will weigh alternatives appropriate to the given situation.
- Students will consider the consequences of those alternatives.
- Students will rank alternatives from best to worst.
- Students will devise a logical argument in favor of their decision.

Vocabulary: business, conserve, ecology, endanger, environment, radar, tourism, violators

Materials: Hand out copies of the story, Predecision Task Sheets, and Decision Sheets.

Lesson Plan Suggestions

- Decide if and when students will discuss the impact environmental groups might have on legislators and legislation.
- Decide if and when students will examine the steps necessary for the passage of bills.
- Decide if and when students will investigate the advantages and disadvantages of tourist communities on the environment,
- Decide how students will be organized into small groups.
- Decide when and how terms will be defined by students.
- Decide which Review and Reflection Questions to ask to focus the general discussion following the story.
- Decide how you will incorporate the decisions from this episode into the next activity you use.

EPISODE 14

Content Focus: Space exploration and past achievements by manned and unmanned vehicles

Episode Title: Off We Go

Duration: 2 class periods

Purpose: To help students reflect upon many scientific accomplishments associated with space exploration

Decision Strategy: Negotiation

Description: The story explores a sample of scientific accomplishments in the area of space exploration. After reading the story, students individually consider fifteen achievements in space travel by completing Predecision Task Sheet 1. Predecision Task Sheet 2 has students consider the importance of these space accomplishments. Once their individual responses are recorded, students meet in groups to discuss their decisions. Then they divide the alternatives into three groups or classes of options. They write a personal decision on the Individual Decision Sheet. Finally, they reach a group consensus and record their decisions on Group Decision Sheets 1 and 2.

Content Objectives

- Students will comprehend 15 accomplishments in the early years of space exploration.
- Students will describe many of the scientific accomplishments associated with space flights with and without humans aboard.

Process Objectives

- Students will make decisions between more and less desirable alternatives.
- Students will group a set of options into three groups: the top, middle, and bottom alternatives.
- Students will consider the consequences of grouping several homogeneous alternatives.
- Students will devise logical reasons in favor of their decision.

Vocabulary: astronaut, commemoration, cosmonaut, launch, orbit, probe, satellite, space probes, space station, suborbital, transatlantic

Materials: Hand out copies of the story, Predecision Task Sheets, and Individual and Group Decision Sheets.

Lesson Plan Suggestions

- Decide if and when students will study scientific accomplishments of unmanned space exploration.
- Decide if and when students will study scientific accomplishments of manned space exploration.
- Decide if and when students will examine in detail the space achievements and disappointments of the former Soviet Union.
- Decide if and when students will examine in detail the space achievements and disappointments of the United States.
- Decide if and when students will examine pictures of space activities and objects.
- Decide when and how terms will be defined by students.
- Decide which Review and Reflection Questions to ask to focus the general discussion following the story.
- Decide how you will help students incorporate the content and concepts from this episode into the next activity you use.

EPISODE 15

Content Focus: Theories on the extinction of dinosaurs

Episode Title: Why? O Why?? O Why???

Duration: 1 class period

Purpose: To help students understand some of the explanations scientists have proposed for the extinction of the dinosaurs

Decision Strategy: Negotiation

Description: This episode has students make individual responses regarding scientists, theories, and theories on the extinction of dinosaurs. After a short reading on the extinction of dinosaurs, students are given a series of nine questions on Predecision Task Sheet 1. After students write answers on their own, put them in small or large groups to discuss their answers with one another. Predecision Task Sheet 2 has students consider the likely accuracy of each theory.

Content Objectives

- Students will describe at least five proposed theories to explain the extinction of dinosaurs.
- Students will consider strengths and weaknesses of each theory as a plausible explanation.
- Students will comprehend widely accepted details about dinosaurs.

Process Objectives

- Students will develop individual responses to information that includes incompatible explanations.
- Students will form reasons for their own responses and answers.
- Students will compare alternative theories.

Vocabulary: asteroid, catastrophe, comet, epidemic, explanation, extinction, species, survival, theory

Materials: Hand out copies of the story, Predecision Task Sheets, and Decision Sheets.

Lesson Plan Suggestions

- Decide if and when students will study other information about various dinosaurs.
- Decide if and when students will explore reference books to get information about why dinosaurs become extinct.
- Decide whether students will complete the Predecision Task Sheets before or after the Decision Sheets.
- Decide if and when films or handouts on dinosaurs will be used for further information on of these prehistoric creatures.
- Decide when and how students will complete the episode only as individuals or will share their responses with others in groups.
- Decide which Review and Reflection Questions to ask to focus the general discussion following the story.
- Decide how students will apply the content and concepts in this episode to the next activity you use.

EPISODE 16

Content Focus: Equipment for science labs

Episode Title: Tough Decisions

Duration: 1 or 2 class periods

Purpose: To help students become familiar with science laboratory equipment

Decision Strategy: Negotiation

Description: The story deals with the selection of science equipment for school laboratories. Individually, students should read the story. If appropriate, they should complete the definitions and descriptions on the optional Predecision Task Sheet. In Predecision Task Sheet 1, students consider the areas in science in which each item might be used. They make individual choices and then meet in small groups to divide the alternatives into three groups or classes. They write their groupings of the items and their reasons for making them on the Individual and Group Decision Sheets.

Content Objectives

- Students will describe 28 different pieces of science equipment.
- Students will describe how different items would be used in science experiments and observations.
- Students will comprehend different areas of science and likely items of scientific equipment used in each.

Process Objectives

- Students will make decisions between more and less desirable alternatives.
- Students will group alternatives into three groups—most important, middle, and least important—according to the priority of each alternative.
- Students will consider the consequences of the losses and gains likely to result from their priority choices.
- Students will formulate logical criteria upon which to base and justify their decisions.

Vocabulary: anatomy, anemometer, aquarium, barometer, chemicals, constellation, dissect, formaldehyde, laboratory, plankton, scalpel, specimens, telescope, terrarium, trajectory

Materials: Hand out copies of the story, Predecision Task Sheets, and Decision Sheets.

Lesson Plan Suggestions

- Decide if and when students will explore areas of science that need specific equipment for experiments.
- Decide if and when students will examine different science equipment catalogues to find additional descriptions of the items.
- Decide if and when students will survey and discuss the science equipment in their own school.
- Decide whether students will complete Predecision Task Sheets 1 and 2 as homework, to be turned in at the start of the episode.
- Decide how students will divide into groups to complete the Group Decision Sheet.
- Decide whether students will complete the optional Predecision Task Sheet.
- Decide which Review and Reflection Questions to ask to focus the general discussion following the story.
- Decide how you will incorporate the concepts from this episode into the next activity you use.

EPISODE 17

Content Focus: Energy conservation in a school setting

Episode Title: We're Running Out of Juice

Duration: 1 class period

Purpose: To help students be aware of ways to conserve energy

Decision Strategy: Negotiation

Description: The story deals with a school's attempt to conserve energy. The Predecision Task Sheet has students individually consider nine energy-savings proposals. Then students meet in groups to divide the alternatives into three groups or classes of options. They record their personal and group choices on the Decision Sheets.

Content Objectives

- Students will consider nine ways energy can be conserved in a school.
- Students will comprehend details about general energy conservation.

Process Objectives

- Students will make decisions between more and less desirable alternatives.
- Students will group alternatives into a set of top, middle, and bottom alternatives.
- Students will consider the consequences of grouping several different alternatives.
- Students will devise logical reasons in favor of their decision.

Vocabulary: caulk, electricity, energy, gasket, insulation, postpone, thermostat, weather-stripping

Materials: Hand out copies of the story, Predecision Task Sheets, and Decision Sheets.

Lesson Plan Suggestions

- Decide if and when students will study and consider possible advantages and disadvantages of alternative energy uses.
- Decide if and when students will complete a cost analysis of energy-saving procedures.
- Decide how students will be organized into groups.
- Decide if and when students will examine graphs of energy use.
- Decide when and how terms will be defined by students.
- Decide which Review and Reflection Questions to ask to focus the general discussion following the story.
- Decide how you will move from this episode and the content it contains to the next activity you use.

EPISODE 18

Content Focus: Occupations in science

Episode Title: Scientists—They're Everywhere!

Duration: 1 or 2 class periods

Purpose: To help students understand the scientific method as it relates to various fields and areas in science

Decision Strategy: Negotiation

Description: The story describes nine occupations in science. Students consider how each scientist may use scientific methods. An explanation of one scientific method is provided. After carefully reading the story, students consider these science occupations. Once their individual responses are recorded, students meet in groups to divide the alternatives into three groups or classes of options.

Content Objectives

- Students will describe nine scientific occupations.
- Students will comprehend one scientific method.
- Students will consider how scientists in nine occupations use scientific methods.

Process Objectives

- Students will make decisions between more and less desirable alternatives.
- Students will group alternatives into a set of top, middle, and bottom alternatives.
- Students will consider the consequences of grouping homogeneous alternatives.
- Students will give logical reasons in favor of their choice.

Vocabulary: anthropologist, archaeologist, astronomer, biologist, botanist, chemist, geologist, historian, hypothesis, meteorologist, oceanographer, scientific method, zoologist

Materials: Hand out copies of the story, Predecision Task Sheets, and Decision Sheets.

Lesson Plan Suggestions

- Decide if and when students will define a "method of science."
- Decide if and when students will interview scientists in different occupations.
- Decide if and when students will outline the nine scientific fields.
- Decide how students will define the vocabulary terms used in the episode.
- Decide which Review and Reflection Questions to ask to focus the general discussion following the story.
- Decide how you will move from this episode to the next activity you use.

EPISODE 19

Content Focus: Botany and zoology as branches of biology

Episode Title: Bo or Zo?

Duration: 1 or 2 class periods

Purpose: To help students understand equivalencies, similarities, and differences between two fields of and occupations of scientists—zoology and botany

Decision Strategy: Invention

Description: The story deals with two students trying to decide which is a "truer" representation of an ideal scientist—a zoologist or a botanist. Areas of specialization in these two branches of biology are included. If the Predecision Task Sheets are to be completed, students should try to do them individually. There is a small-group discussion section in which students are asked to find out what other students think about the problem and to clarify terms before individually responding to the Individual Decision Sheet. Then students complete the Group Decision Sheet.

Note: In modern scientific thinking, there is no such thing as a "true" scientist or area of science. Rather, individuals or areas are considered scientific according to their adherence to standards associated with systematic methods of investigation, data collection, measurement, and interpretive perspective. This activity mentions "true" and "truer" because many people believe one area of science is the "truer" one. Teachers should do all they can to dispel the misconception of a one "true" science.

Content Objectives

- Students will describe areas of equivalence, similarity, and difference between zoology and botany.
- Students will comprehend areas of specialization for zoologists and botanists.
- Students will define terms related to zoology and biology.
- Students will describe what they mean when they use the term *scientist.*

Process Objectives

- Students will develop on their own a series of alternatives appropriate for a given situation.
- Students will weigh the consequences of the alternatives they have identified as being appropriate to this situation.
- Students will examine the problem from different points of view.

Vocabulary: anatomy, bacteria, biology, botanist, chlorophyll, crossbreeding, cytologist, ecologist, fossil, geneticist, nucleus, pathologist, primitive, taxonomist, zoologist

Materials: Hand out copies of the story, Predecision Task Sheets, and Decision Sheets.

Lesson Plan Suggestions

- Decide if and when students will study information relative to the two branches of biology.
- Decide if and when students will examine areas of specialization in zoology and botany.
- Decide how students will be organized into small groups to share information concerning the Predecision Task Sheets.
- Decide when and how terms will be defined by students.
- Decide if and when students will examine branches of other areas of science.
- Decide which Review and Reflection Questions to ask to focus the general discussion following the story.
- Decide how you will move from this episode and its content to the next activity you use.

EPISODE 21

Content Focus: Relationships among plants, animals, and the environment

Episode Title: A Harey Situation

Duration: 1 class period

Purpose: To help students understand how personal conflicts between two things one believes in can arise

Decision Strategy: Invention

Description: The story deals with a boy who likes animals and his reactions during a rabbit-hunting trip. The destruction to local farmland caused by these animal pests is pointed out. Each student should carefully read the story and make the decision each believes is best. Then, as time allows, students meet in small groups to look at the problem from different points of view and devise a group consensus solution.

Content Objectives

- Students will understand some duties of a veterinarian's assistant.
- Students will realize the need for control of certain species for environmental protection.
- Students will understand how hunting animals may be one form of protecting the ecological balance of nature.

Process Objectives

- Students will develop on their own a series of alternatives that are appropriate for a given situation.
- Students will weight the consequences of those alternatives.
- Students will examine a problem from different points of view.

Vocabulary: ecology, environment, hare, pest, predator, species control, sterilize, veterinarian

Materials: Hand out copies of the story, Predecision Task Sheets, and Decision Sheets.

Lesson Plan Suggestions

- Decide if and when students will discuss the role of a veterinarian.
- Decide if and when students will examine information on hunting regulations, procedures, and safety.
- Decide if and when students will study the differences and relationships between rabbits and hares and their respective eating habits.
- Decide if and when students will discuss the idea of species control for the betterment or detriment of the environment.
- Decide if and when students will study agricultural pests (animals or insects) and attempts to control and eliminate them.
- Decide if and when students will examine species control methods used in their own homes and communities (for example, bug sprays).
- Decide how students will be organized into small groups to share their ideas.
- Decide which Review and Reflection Questions to ask to focus the general discussion following the story.
- Decide how you will help students apply the content and concepts from this episode to the next activity you use.

EPISODE 22

Content Focus: Possible consequences of a nuclear power plant crisis

Episode Title: Whose Fault?

Duration: 1 or 2 class periods

Purpose: To help students understand dilemmas involved in finding data that may impact on the use of technology, such as the building of a nuclear power plant

Decision Strategy: Invention

Description: The story deals with a fault being discovered by a member of a construction crew at a nuclear power plant site. Predecision Task Sheet 1 has students review the information found in the story. Predecision Task Sheet 2 has them consider how the construction of the nuclear plant may affect various people. Students will record their individual decision on the Individual Decision Sheet, then meet in groups to discuss the topic and to write a group consensus decision on the Group Decision Sheet.

Content Objectives

- Students will define terms related to nuclear energy and fault lines.
- Students will consider alternative energy sources.
- Students will understand some of the problems encountered by observant skilled investigators.
- Students will know the names of some of the buildings that make up a nuclear power plant facility.

Process Objectives

- Students will consider the consequences of the options given to them.
- Students will invent and support reasons to justify their decision.

Vocabulary: after bay reservoir, conglomerate, containment building, faulting, geothermal power, nuclear energy, pressurized water reactor

Materials: Hand out copies of the story, Predecision Task Sheets, and Decision Sheets.

Lesson Plan Suggestions

- Decide if and when students will examine the U.S. Atomic Energy Commission guidelines for the construction of a nuclear power plant.
- Decide if and when students will review the geological and topological changes that take place along fault lines.
- Decide if and when students will examine the possible dangers of faulting under a nuclear power plant site.
- Decide how students will be organized into small groups to share their ideas.
- Decide if and when students will study a diagram of a pressurized water reactor system.
- Decide if and when students will debate the pros and cons of nuclear energy.
- Decide how and when the terms used in the episode will be defined.
- Decide which Review and Reflection Questions to ask to focus the general discussion following the story.
- Decide how you will help students move from this episode to the next activity you use.

EPISODE 24

Content Focus: Noise, music, and noise pollution

Episode Title: Hear Ye! Hear Ye!

Duration: 1 or 2 class periods

Purpose: To help students understand the effects of noise on humans and the environment

Decision Strategy: Invention

Description: The story deals with a loud rock group hired for a school dance. The Predecision Task Sheets provide information on decibel levels of various noises and the likely consequences of such sounds on human beings. Each student should complete the Predecision Task Sheets before going on to the Decision Sheets. Students are encouraged to reach a group consensus decision following their individual decisions.

Content Objectives

- Students will comprehend the meaning of a *decibel level* of sound.
- Students will consider the need for control of certain noises.
- Students will analyze the effects of various types of sound on human beings.
- Students will define *noise pollution* and *noise.*

Process Objectives

- Students will develop on their own a series of alternatives for a given situation.
- Students will weigh the consequences of those alternatives.
- Students will examine a problem from different points of view.

Vocabulary: decibel, hearing, jackhammer, noise, pneumatic drill, pollution, sonic boom, sound

Materials: Hand out copies of the story, Predecision Task Sheets, and Decision Sheets.

Lesson Plan Suggestions

- Decide if and when students will examine types of noise pollution.
- Decide when students might discuss how much exposure to loud music is necessary to result in permanent hearing loss.
- Decide if and when students will discuss how music, noise, and pollution are alike and different.
- Decide whether students will define *decibel, noise, music,* and *noise pollution.*
- Decide if and when students will visit a hearing-aid center or have a hearing specialist visit the class.
- Decide which Review and Reflection Questions to ask to focus the general discussion following the story.
- Decide how you will move from this episode to the next activity you use.

EPISODE 28

Content Focus: Physiology, health, and athlete performance

Episode Title: Athletes' Feats

Duration: 1 class period

Purpose: Students will understand the importance of appropriate training prior to an athletic event

Decision Strategy: Exploration

Description: The story deals with two friends who compete in long-distance races. Several marathon training methods are described. Each student should carefully read the story and write responses to the questions before whole-class discussion begins.

Content Objectives

- Students will evaluate the decisions competitive athletes must make about their own training and health.
- Student will understand several marathon-training methods.

Process Objectives

- Students will reflect upon the application of relevant information to making a decision.
- Students will consider and develop individual reasons for making important decisions.
- Students will invent personal reactions to decisions that were made.

Vocabulary: carbohydrate, competitor, concentrate, diligent, marathon, pace, rival

Materials: Hand out copies of the story and questions to each student.

Lesson Plan Objectives

- Decide if and when students will study and consider the importance of training prior to a marathon.
- Decide if and when students will examine special dieting and fasting methods used by marathon competitors.
- Decide if and when students will read about several famous long-distance runners and their training programs.
- Decide when and how students will define terms used in the story.
- Decide which Review and Reflection Questions to ask to focus the general discussion following the story.
- Decide how you will help students move from this activity to the next activity you use.

EPISODE 30

Content Focus: Restricting the importation of injurious wildlife

Episode Title: Ye May Keep Your Pet Rocks

Duration: 1 class period

Purpose: To help students understand the need for import regulations of injurious wildlife

Decision Strategy: Exploration

Description: The reading addresses the U.S. Department of the Interior's announcement of proposed regulations controlling the importation of injurious wildlife. Examples of injuries received from imported animals are cited. Affects of the regulations on the pet industry are pointed out. Each student should carefully read the story and write responses to the questions before the large-group discussion begins.

Content Objectives

- Students will evaluate the results of the proposed regulations on the pet industry.
- Students will recall the names of a number of imported animals.
- Students will comprehend possible injuries and diseases that could be caused by some imported animals.
- Students will define terms used in the episode.

Process Objectives

- Students will consider and develop individual solutions to the problem they examine.
- Students will invent and reflect upon personal reasons for the decisions they must make.
- Students will consider likely consequences of a number of possible solutions that could be adopted.

Vocabulary: amphibian, exotic, hepatitis, horticulture, inflamed, nervous system, Newcastle disease, reptile, respiratory system, salmonella poisoning, tuberculosis, venomous

Materials: Hand out copies of the reading and Decision Sheet to each student.

Lesson Plan Suggestion

- Decide if and when students will know examples of imported animals.
- Decide if and when students will investigate the long-range effects of the diseases mentioned in the story.
- Decide if and when students will conduct a survey of local pet stores to gather data on imported species.
- Decide if and when students will examine the list of "low risk" wildlife.
- Decide if and when students will investigate how to obtain a permit to import animals.
- Decide when and how terms will be defined by students.
- Decide which Review and Reflection Questions to ask to focus the general discussion following the story.
- Decide how you will help students incorporate the content and concepts from this episode into the next activity you use.

REFERENCES

Chapter 1

Arons, A. "Achieving Wider Scientific Literacy." *Daedulus,* 112(1) (1983), pp. 91–122.

Bragaw, D.H. "Society, Technology, and Science: Is There Room for Another Imperative?" *Theory into Practice,* 31(1) (1992), pp. 4–12.

Casteel, J.D., and R.E. Yager. "Science as Mind-Effected and Mind-Effecting Inquiry." *Journal of Research in Science Teaching,* 4(3) (1966), pp. 127–136.

Duschl, R. *Restructuring Science Education: The Importance of Theories and Their Development.* New York: Teachers College Press, 1990.

Forbes, R.J. *The Conquest of Nature: Technology and Its Consequences.* New York: Frederick A. Praeger, 1968.

Gradwell, J.B. "Twenty Years of Technology Education." *The Technology Teacher,* 47 (1988), pp. 30–33.

Heron, M. "The Nature of Scientific Inquiry." *School Review,* 79 (1970), pp. 171–212.

Hetzler, S.A. *Technological Growth and Social Change.* New York: Frederick A. Praeger, 1969.

Hofstein, A., G. Aikenhead, and K. Riquarts. "Discussions over STS at the Fourth IOSTE Symposium." *International Journal of Science Education,* 10(4) (1988), pp. 357–366.

Karian, M.E. "Verification of Instructional Objectives and a Sample Unit in Technology Education Addressing Environmental Impacts." Doctoral dissertation, Arizona State University, 1991.

Klopher, L. "The Teaching of Science and the History of Science." *Journal of Research in Science Teaching,* 6 (1969), pp. 87–95.

Kranzberg, M., and C.W. Pursell. *Technology in Western Civilization,* Vol. 1. New York: Oxford University, 1967.

Maley, D. "Technology Literacy as a Major Thrust for Technology Education." *Journal of Epsilon Pi Tau,* 13(1) (1987), pp. 44–49.

Markert, L.R. *Contemporary Technology.* South Holland, Ill.: Goodheart-Willcox, 1989.

National Council for the Social Studies. "Teaching Science-Related Social Issues: A Position Paper." Washington, D.C., 1982.

National Science Board Commission on Precollege Education in Mathematics, Science and Technology. "Educating Americans for the 21st Century: A Report to the American People and the National Science Board." Washington, D.C.: National Science Foundation, 1983.

Ogburn, W. "Technology as Environment." *Sociology and Social Research,* 16 (1966), pp. 3–9.

Pytlik, E.C., D.P. Lauda, and D.L. Johnson. *Technology, Change and Society.* Worcester, Mass.: Davis, 1985.

Savage, E. and L. Sterry, eds. *A Conceptual Framework for Technology Education.* Reston, Va.: International Technology Education Association, 1991.

Toulmin, S. *The Philosophy of Science: An Introduction.* New York: Harper and Row, 1960.

Chapter 3

Casteel, J.D. *Learning to Think and Choose.* Santa Monica, Calif.: Goodyear, 1978.

Casteel, J.D., and R.J. Stahl. "Doorways to Decision-Making: A Handbook for Designing Academic Decision-Making Activities." Unpublished manuscript, 1993.

Casteel, J.D., and R.E. Yager. "Science as Mind-Effected and Mind-Effecting Inquiry." *Journal of Research in Science Teaching,* 4(3) (1966), pp. 127–136.

Hunt, B.S. "Effects of Values Activities on Content Retention and Attitudes of Students in Junior High Social Studies Classes." Ph.D. dissertation, Arizona State University, 1981.

Stahl, R.J. "Achieving Values and Content Objectives Simultaneously Within Subject Matter Oriented Social Studies Classrooms." *Social Education,* 45(7) (1981), pp. 580–585.

______. "Working with Values and Moral Issues in Content-Centered Science Classrooms." *Science Education,* 63(2) (Apr. 1979), pp. 183–194.

Chapter 4

Casteel, J.D., and R.J. Stahl. "Doorways to Decision-Making: A Handbook for Designing Academic Decision-Making Activities." Unpublished manuscript, 1993.

______. *Value Clarification in the Classroom: A Primer.* Santa Monica, Calif.: Goodyear, 1975.

Johnson, D.W., R.T. Johnson, and E.J. Holubec. *Circles of Learning: Cooperation in the Classroom,* 4th ed. Edina, Minn.: Interaction Book Company, 1993.

Stahl, R.J. *A Blueprint for Designing Structured Group Decision-Making Episodes: An Introductory Module.* Tempe, Ariz.: Arizona State University, 1979.

______. "The Casteel-Stahl Model as An Alternative Cooperative Learning Strategy." Workshop conducted at Curtin University of Technology, Perth, Western Australia, June 1987.

______. "A Context for 'Higher-Order Knowledge': An Information-Constructivist (IC) Perspective with Implications for Curriculum and Instruction." *Journal of Structural Learning,* 11(3) (1992), pp. 189–218.

______. Using Small Group Decision-Making Activities to Achieve a Cooperative Learning Classroom: A Model and Examples." Workshop presented at the Fall Conference of the Arizona Council for the Social Studies, Oct. 1990.

Stahl, R.J., ed. *Cooperative Learning in Language Arts: A Handbook for Teachers.* Menlo Park, Calif.: Addison-Wesley, 1995.

______. *Cooperative Learning in Science: A Handbook for Teachers.* Menlo Park, Calif.: Addison-Wesley, 1995.

______. *Cooperative Learning in Social Studies: A Handbook for Teachers.* Menlo Park, Calif.: Addison-Wesley, 1994.

Stahl, R.J., and R.L. VanSickle, eds. *Cooperative Learning in the Social Studies Classroom: An Invitation to Social Study.* Bulletin No. 87. Washington, D.C.: National Council for the Social Studies, 1992.

Appendix

Crichton, M. *Jurassic Park.* New York: Ballantine Books, 1990.

Dixon, D. *Dinosaurs.* Honesdale, Penn.: Boyds Mills Press, 1993.